101 SCIENCE SURPRISES

Exciting Experiments with Everyday Materials

ROY RICHARDS

Illustrated by Alex Pang

Sterling Publishing Co., Inc.   New York

CONTENTS

3 On Pattern

For Phoebe Jane

Library of Congress Cataloging-in-Publication Data

Richards, Roy.
    101 Science surprises : exciting experiments with everyday
materials / by Roy Richards : illustrated by Alex Pang.
        p.   cm.
    Originally published: London : Simon & Schuster, c1992.
    Includes index.
    Summary: Presents experiments and activities involving natural
observation, geometric patterns, simple chemistry, and more.
    ISBN 0–8069–8822–3
    1. Science—Experiments—Juvenile literature.   2. Scientific
recreations—Juvenile literature.   [1. Science—Experiments.
2. Experiments.   3. Scientific recreations.]   I. Pang, Alex, ill.
II. Title   III. Title: One hundred one science surprises.
IV. Title: One hundred and one science surprises.
Q164.R53 1992   92–32491
507.8—dc20   CIP
                                                          AC

10 9 8 7 6 5 4 3 2 1

First paperback edition published in 1994 by
Sterling Publishing Company, Inc.
Originally published in Great Britain by
Simon & Schuster Young Books under the title
101 Science Surprises © 1992 by Roy Richards
Illustrations © 1992 by Alex Pang
Distributed in Canada by Sterling Publishing
c/o Canadian Manda Group, P.O. Box 920, Station U
Toronto, Ontario, Canada M8Z 5P9
Printed and bound in Hong Kong
All rights reserved

Sterling ISBN 0–8069–8822–3 Trade
              0–8069–8823–1 Paper

1
on Line

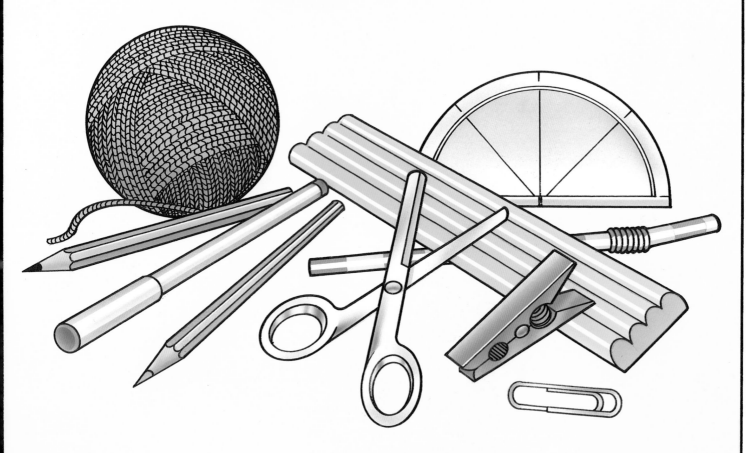

INTRODUCTION

Have you ever seen a snail moving over paving stones? It leaves a line behind it, a slime trail. Lots of animals follow a trail. You do every day as you go from home to school. Have you ever been in a maze? You have to follow the right line to get to the center, and you have to find the right line to get out. In "On Line" you'll find all sorts of things to make and do using lines. You can cut a string of figures, weigh a letter by stretching a rubber band, make a diver that goes up and down, play tunes, send messages, fire a coin or a soccer ball, weave things in and out, and create wonderful patterns. You could use lines to decorate book covers, make intriguing greetings cards or design a wallpaper pattern. A list of all the materials you need is given on page 101.

SNAIL WAYS

Snails leave a line of slime behind them as they move. This slime helps them to move more easily on rough surfaces. It is also sticky and helps them cling to things. You can make a record of this line.

1 Cut the bottom off a cardboard box to make a tray.

2 Line the bottom of the tray with black paper or paint it black. If you paint it leave the paint to dry.

3 Put a snail in the tray.

4 Wrap the tray round with plastic wrap pricked with holes to let in air.

5 Leave the tray overnight.

6 In the morning you will see a white slime trail left by the snail.

7 Put a piece of string carefully along the trail to match its path and its length.

Cut the string and measure its length. It will show you how far the snail moved in the night.

Let the snail go again!

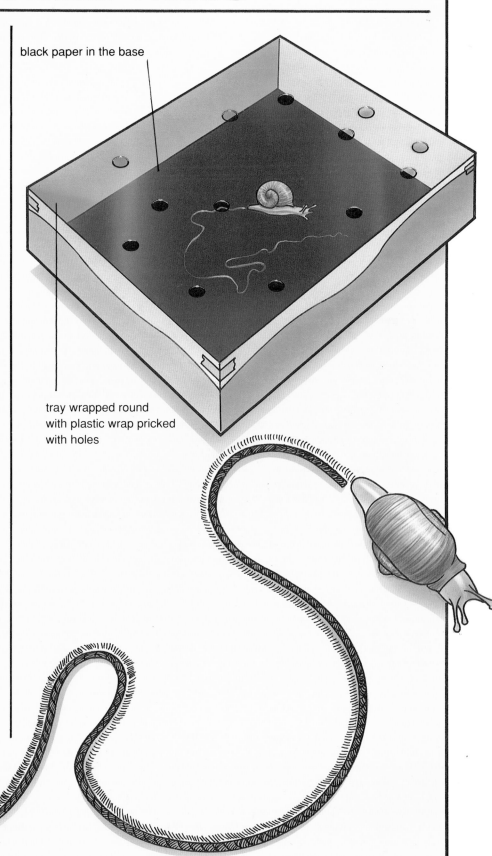

black paper in the base

tray wrapped round with plastic wrap pricked with holes

MAZES

According to Greek legend there was a fierce monster called the Minotaur. He lived in a maze, or a labyrinth as it is sometimes called. A brave Greek called Theseus found his way into the labyrinth and killed the Minotaur. Theseus was clever enough to find his way out again.

This is a plan of the labyrinth.
Can you find your way to the center and out again?

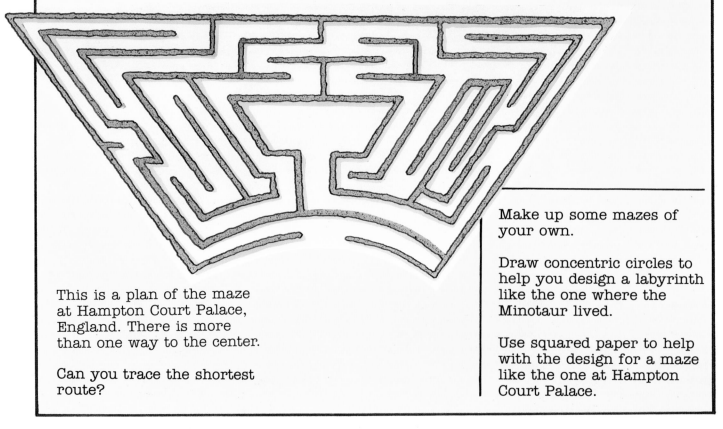

This is a plan of the maze at Hampton Court Palace, England. There is more than one way to the center.

Can you trace the shortest route?

Make up some mazes of your own.

Draw concentric circles to help you design a labyrinth like the one where the Minotaur lived.

Use squared paper to help with the design for a maze like the one at Hampton Court Palace.

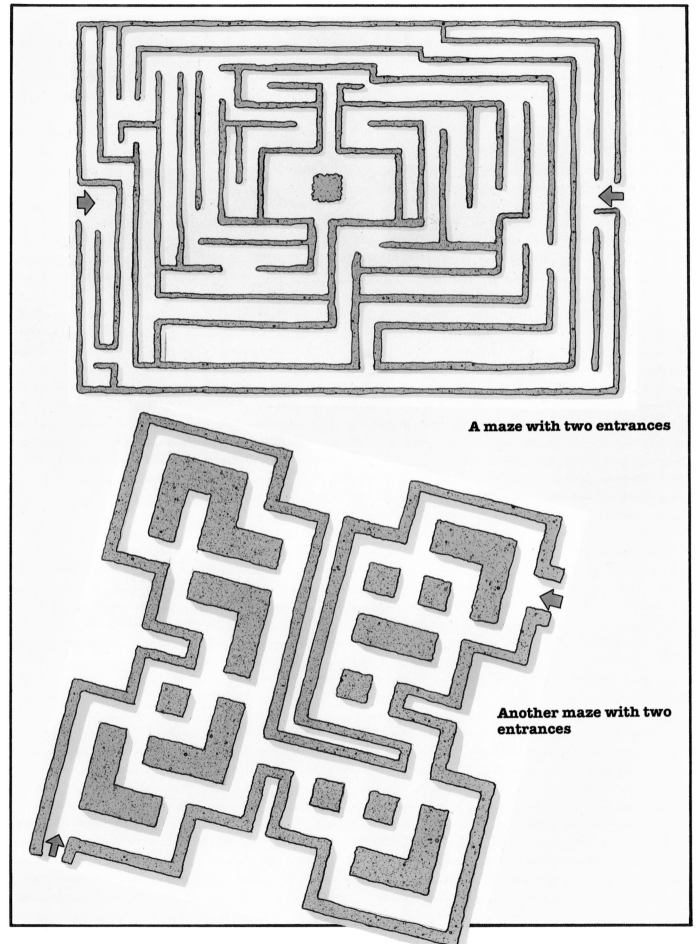

A maze with two entrances

Another maze with two entrances

A MARBLE MAZE

The marble maze will test the steadiness of your hand.

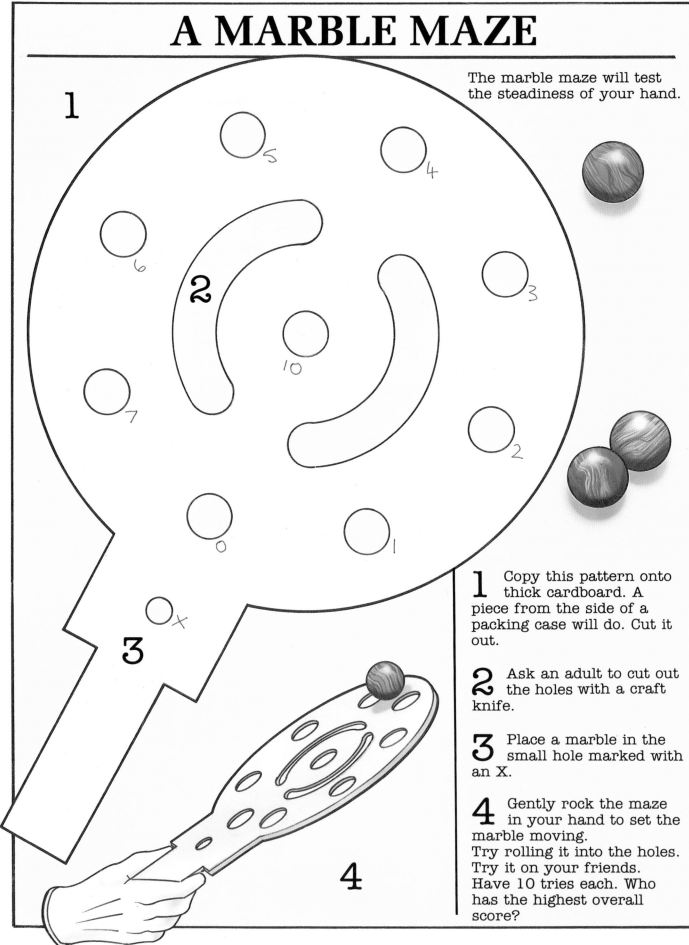

1 Copy this pattern onto thick cardboard. A piece from the side of a packing case will do. Cut it out.

2 Ask an adult to cut out the holes with a craft knife.

3 Place a marble in the small hole marked with an X.

4 Gently rock the maze in your hand to set the marble moving. Try rolling it into the holes. Try it on your friends. Have 10 tries each. Who has the highest overall score?

A FEELY MAZE

Can your friends find their way through a maze blindfolded? Make up a touch maze to test them.

1 Gather together a heavy piece of cardboard (8 in. by 8 in. is a good size), some drinking straws, a pack of graded sandpaper, glue and scissors.

2 Draw a pathway on the card. It can be similar to the one shown below but make your own design.

3 Stick drinking straws to the sides of the pathway. Cut straws to size where necessary.

4 Use the graded sandpaper to give instructions on which way to turn. Rough means turn right, medium means turn left, fine means go straight ahead. Stick a small square of the appropriate grade at each bend.

5 Test your friends. Can they find their way through by touch alone? Each must only use a forefinger. No peeking!

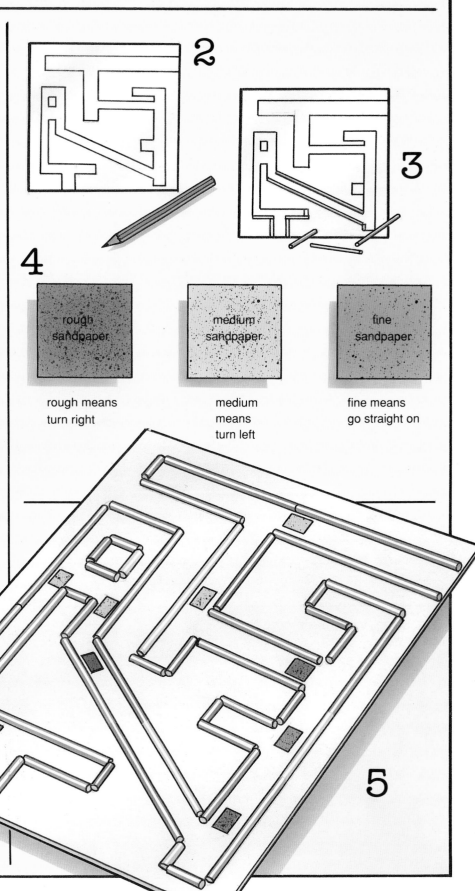

rough means turn right

medium means turn left

fine means go straight on

CUTOUTS

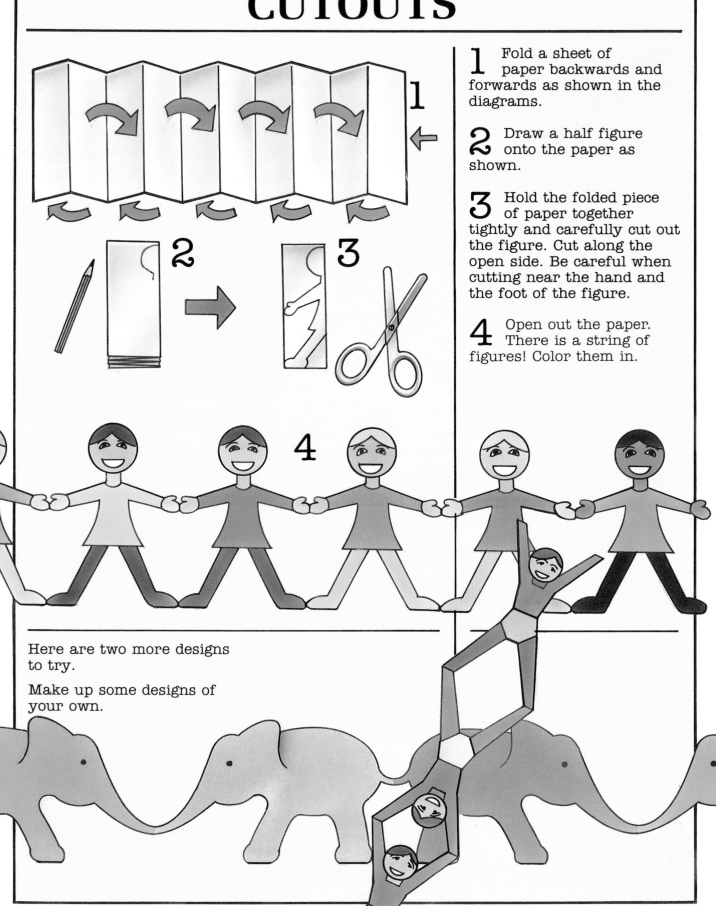

1 Fold a sheet of paper backwards and forwards as shown in the diagrams.

2 Draw a half figure onto the paper as shown.

3 Hold the folded piece of paper together tightly and carefully cut out the figure. Cut along the open side. Be careful when cutting near the hand and the foot of the figure.

4 Open out the paper. There is a string of figures! Color them in.

Here are two more designs to try.

Make up some designs of your own.

A LETTER BALANCE

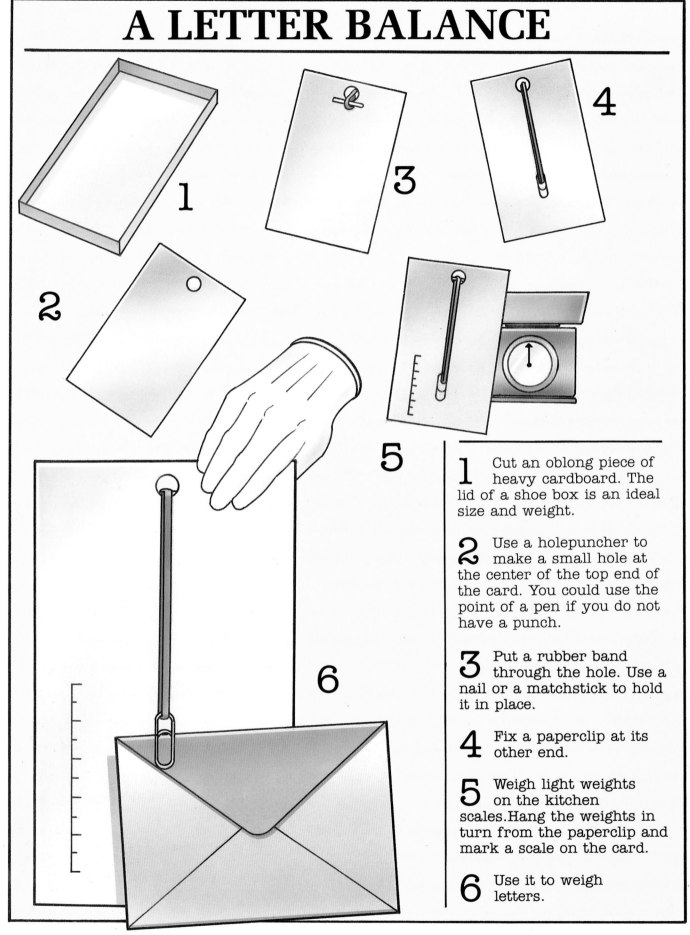

1 Cut an oblong piece of heavy cardboard. The lid of a shoe box is an ideal size and weight.

2 Use a holepuncher to make a small hole at the center of the top end of the card. You could use the point of a pen if you do not have a punch.

3 Put a rubber band through the hole. Use a nail or a matchstick to hold it in place.

4 Fix a paperclip at its other end.

5 Weigh light weights on the kitchen scales. Hang the weights in turn from the paperclip and mark a scale on the card.

6 Use it to weigh letters.

MUSICAL INSTRUMENTS

Here are two musical instruments to make. For the first instrument you need a board, two tongue depressers, two screw-eyes and some nylon fishing line.

1 Ask an adult to cut you a piece of board as wide as the length of your tongue depresser.

2 Ask the adult to make two saw-cuts as shown.

3 Wedge a depresser into each saw-cut.

4 Tap each depresser gently but firmly with a hammer to make sure it is securely in place.

5 Put in two screw eyes, one at either end of the board.

6 Tie a piece of nylon fishing line between the two screw eyes. Make sure you tie it as tightly as possible. Pluck the line to get a note.

For the second instrument you need an old plastic bucket, a button, a broom handle and some nylon fishing line.

1 Ask an adult to make a small hole at the center of the bottom of the bucket for you.

2 Tie one end of the nylon fishing line to a button that is bigger than the hole in your bucket. Thread the other end through the hole and tie it firmly to the end of the broom handle.

3 Place the other end of the broom handle on the base of the upturned bucket. Pull it slightly outward to draw the fishing line taut. Use your foot to keep the bucket on the ground. Pluck the line to get a note. Slackening and tightening the line as you pluck will give different notes.

A TIN-CAN TELEPHONE

You will need two thin tin cans, a hammer, a nail, two matchsticks and some string.

1 Ask an adult to use the hammer and nail to make a hole in the base of each can.

2 Draw the string through the holes in the cans and tie each end to a matchstick.

3 Ask someone else to hold one can while you hold the other. Draw the string really tightly between you.

4 Speak quietly into the can to send messages to a friend. Don't shout. If the string is held tight the vibrations will carry well.

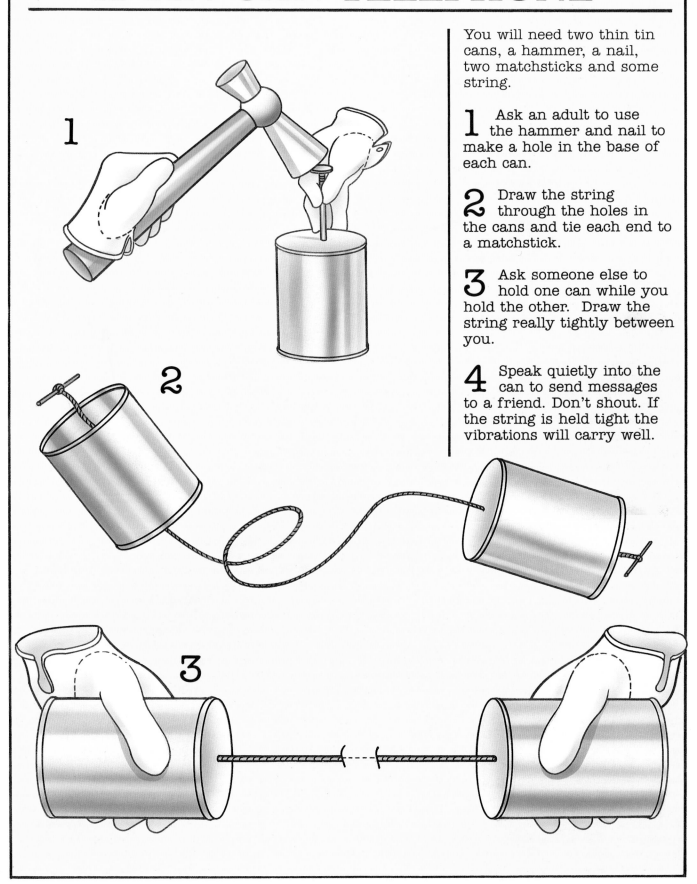

A CARTESIAN DIVER

1 Fill a plastic bottle with water right to the top.

2 Make a diver from a drinking straw, paperclip and Plasticine as shown.

3 Test that it floats in a glass of water. Add Plasticine until it *only just* floats.

4 Once the diver is just floating, put it in the bottle of water.

5 Screw the top on the bottle.

6 Squeeze the bottle gently. The diver should sink. Stop squeezing the bottle and it should rise. A diver like this is called a Cartesian diver.

Remember, you must get the diver only just floating to begin with.

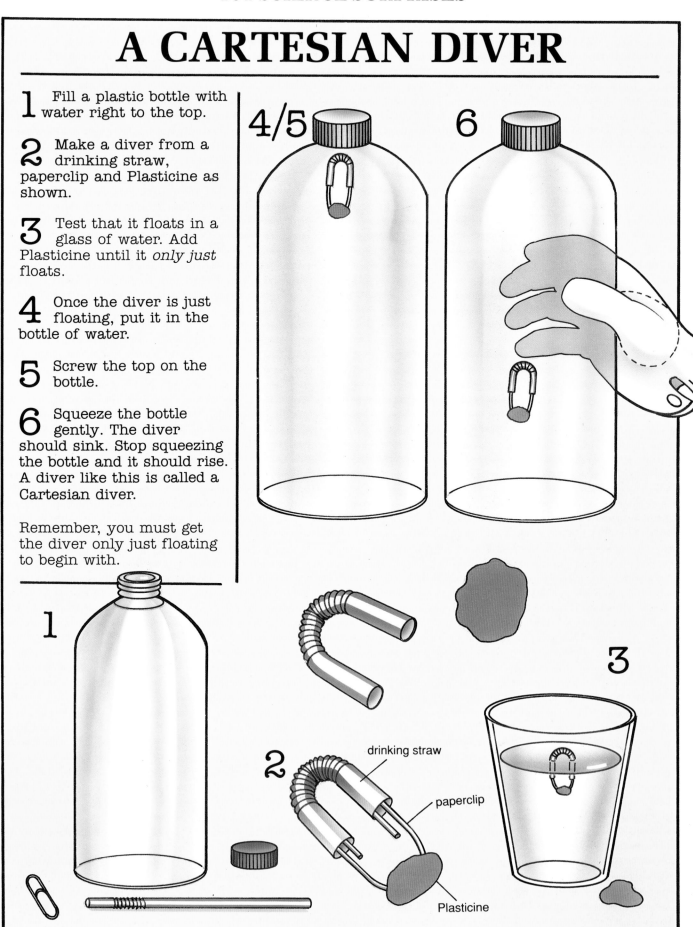

drinking straw

paperclip

Plasticine

BLOTTING PAPER BEASTS

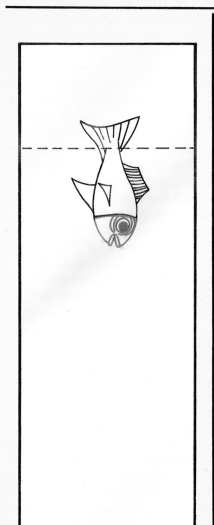

1 Cut a piece of white blotting paper 2 in. by 6 in. Draw a dotted line as shown in pencil. Sketch a fish in pencil.

2 Draw the head of the fish in red felt-tip pen.

3 Fold the top of the strip along the dotted line.

4 Hook the fold over the top of a jam jar full of water. Watch what happens as the water soaks down the blotter.

5 Try other colors.

MORE BEASTS

1 Cut two strips of white blotting paper each 2 in. by 12 in.

2 Draw a snail in pencil on each.

3 Put a thick red line with a felt-tip pen in front of one snail, and a thick blue line in front of the other.

4 Hold each in a saucer of water. As the water soaks along see which 'snail' wins the race.

Have competitions with your friends.

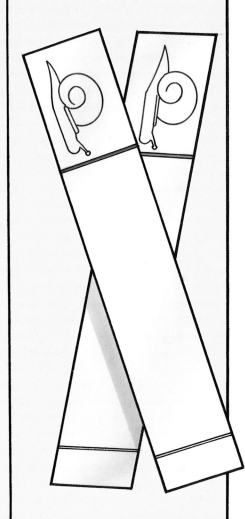

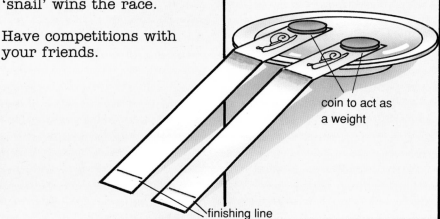

coin to act as a weight

finishing line

A CLOTHESPIN SHOOTER

To make this great shooter you will need a strong clothespin and a rubber band.

1 Pull the spring up and thread the rubber band under the wire on either side of the peg.

2 Push a small paper pellet against the elastic and into the jaws of the clothespin.

3 Close the clothespin.

4 When you squeeze the clothespin open, the rubber band will shoot the pellet forward.

Make up some shooting games.

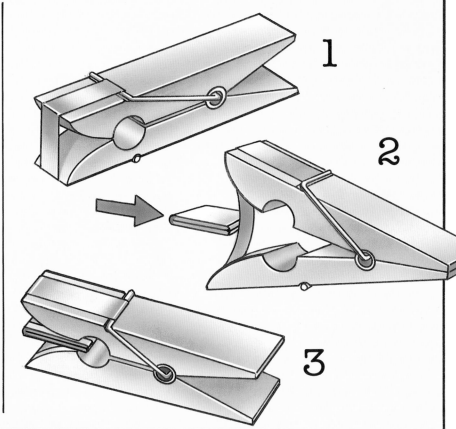

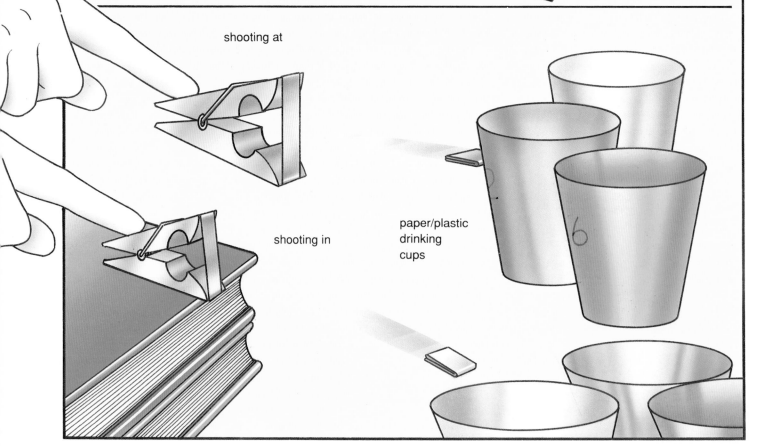

shooting at

shooting in

paper/plastic drinking cups

A CLOTHESPIN SOCCER PLAYER

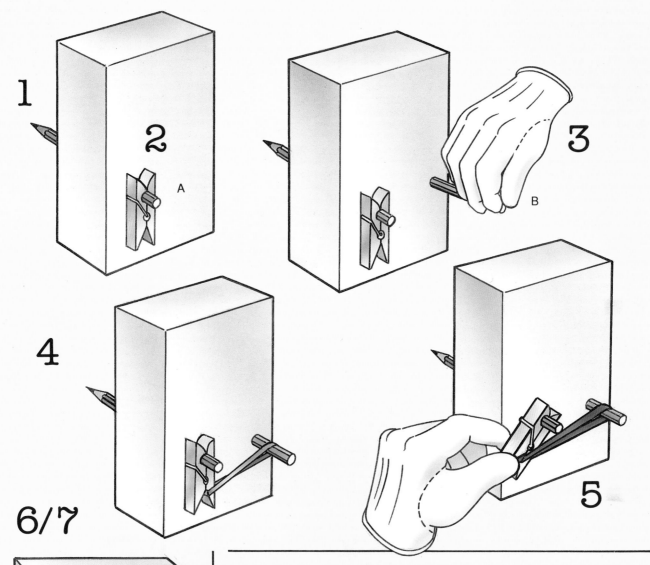

You need a clothespin, a block of polystyrene, a rubber band, two pencils, and a marble.

1 Push a pencil (A) firmly into the block of polystyrene.

2 Hang the clothespin from it. It should move easily on the skewer.

3 Push in the other pencil (B).

4 Put the rubber band over the end of the clothespin and around the pencil as shown.

5 Pull back the clothespin. If you have spaced your two pencils well you should get a good stretch in the rubber band.

6 Release the peg to strike the marble.

7 Decorate the polystyrene block using felt-tip pens.

PAPER WEAVING

1 Fold a sheet of paper in half and cut slits as shown.

2 Open it out and weave in colored strips of paper.

3 Weave different patterns.

plain

twill

satin

Another way to weave is just by using strips of paper. Interweave strips of one color with strips of another color.

Put card frames around your finished weaving and display the patterns around your home.

WOOL WEAVING

Make your own woolen material using wool and a piece of cardboard.

1 Cut notches in a piece of thick cardboard to make your loom.

2 Wind the wool fairly tightly around the card. Hold each end in place with tape.

3 With a large darning needle weave a different colored wool through the wool on the card.

4 Cut the loops of wool which go through the cardboard notches. The illustration is drawn large to show you how to do it. Put your notches closer together and use more wool in your work to get a tighter weave.

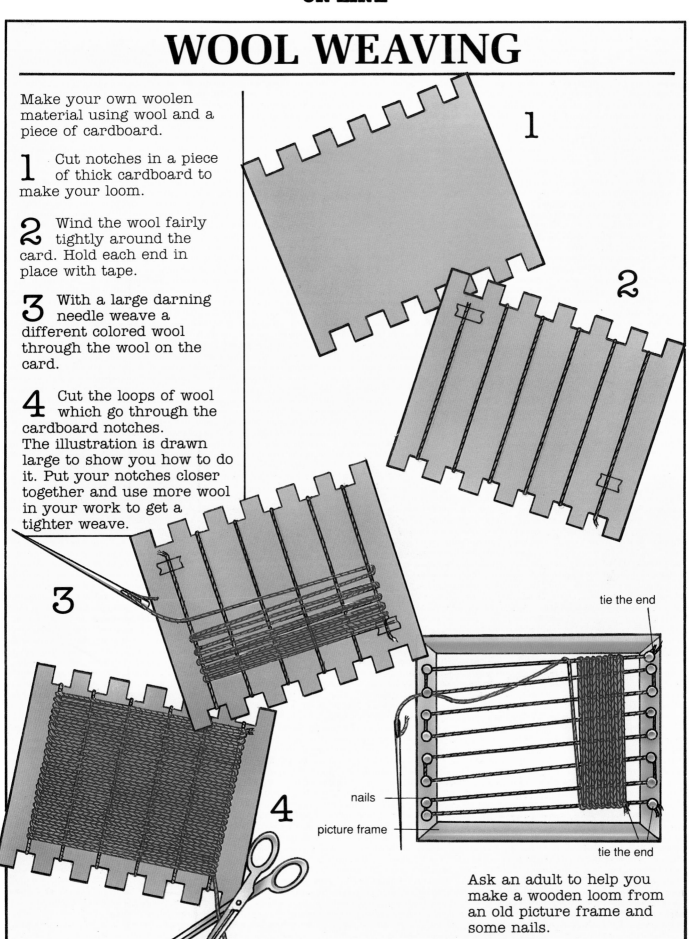

tie the end

nails

picture frame

tie the end

Ask an adult to help you make a wooden loom from an old picture frame and some nails.

A BRIDGET CROSS

Here is a simple form of weaving using straw stems.

1 Make a central cross from two thick pieces of straw.

2 Start weaving like this with a single piece of straw. This straw is called the weaver straw.

3 Take the weaver straw under A and over the center to bend down behind B.

4 Wind the straw back across the center. Bend it around the back of C over the front and up behind D.

5 Add new weaver straws as you go along by inserting the thin end of the new straw into the thick end of the old weaver straw.

6 Continue until the cross is finished. Tie off the final end with a piece of thread. Cut four heads of wheat, leaving a stem on each one. Insert the stems in the corners of the cross.

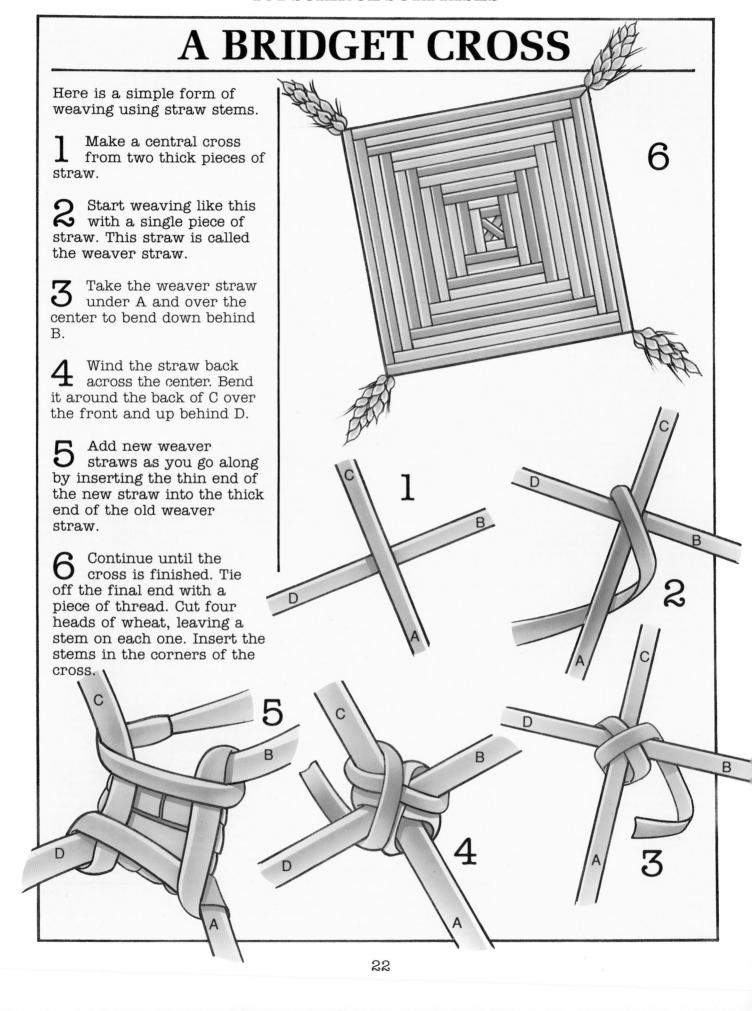

A MÖBIUS STRIP

A Swede called August Ferdinand Möbius discovered that it was possible to have a one-sided piece of paper. Some strange things happen to it. You try. Gummed paper is easiest to use.

Use strips of paper 12 in. by $1\frac{1}{4}$ in.

1 Make two loops. Put a half twist in one.

2 Cut your two loops in half lengthwise. What happens? Is it the same in each case?

Make another loop with a half twist. Try cutting it $\frac{1}{3}$ of the way from the edge.

Make another loop with a half twist. Try cutting it $\frac{1}{4}$ of the way from the edge.

Make another loop with a half twist. Try cutting it $\frac{1}{5}$ of the way from the edge.

If it is difficult use wider strips.

Now make a double loop. To create your double loop you need to make your original two loops, one plain and one with a half twist. Join them together at right angles with glue as shown on the right. Cut both lengthwise.
What happens?

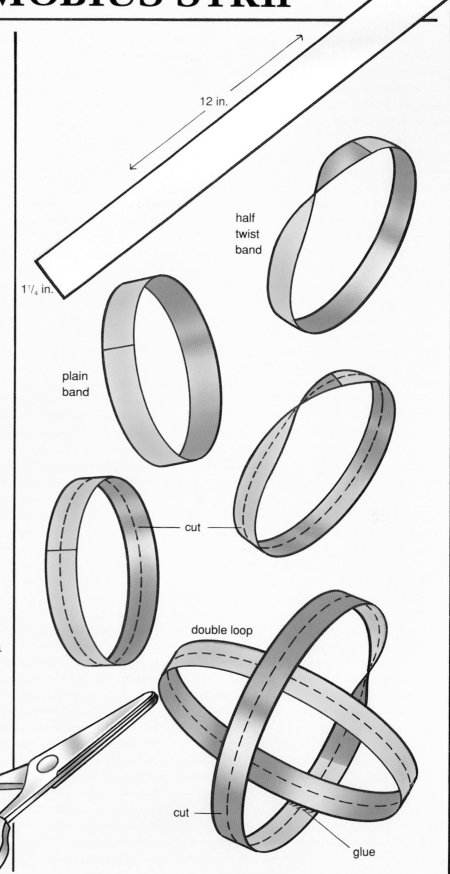

12 in.

$1\frac{1}{4}$ in.

half twist band

plain band

cut

double loop

cut

glue

PATTERNS FROM TRIANGLES

Draw a triangle.

1 First draw the base line A B.

2 Set a compass with the distance between point and pencil equal to the length of the base line.

3 Placing the point of the compass on first A and then B, draw arcs C D and E F.

4 Join the point where the arcs cross to each end of the base line. You now have an equilateral triangle. That is to say all the sides are the same length.

5 Make designs in the triangles.

A design made by quartering each side and drawing parallel lines

A design made by joining the midpoints of each side

A design made by working from the midpoints of each side

PATTERNS FROM SQUARES

1 Draw a square. Make sure each side is the same length and that each angle is 90°.

2 Make designs in the square.

Can you work out how to make the two designs shown below?

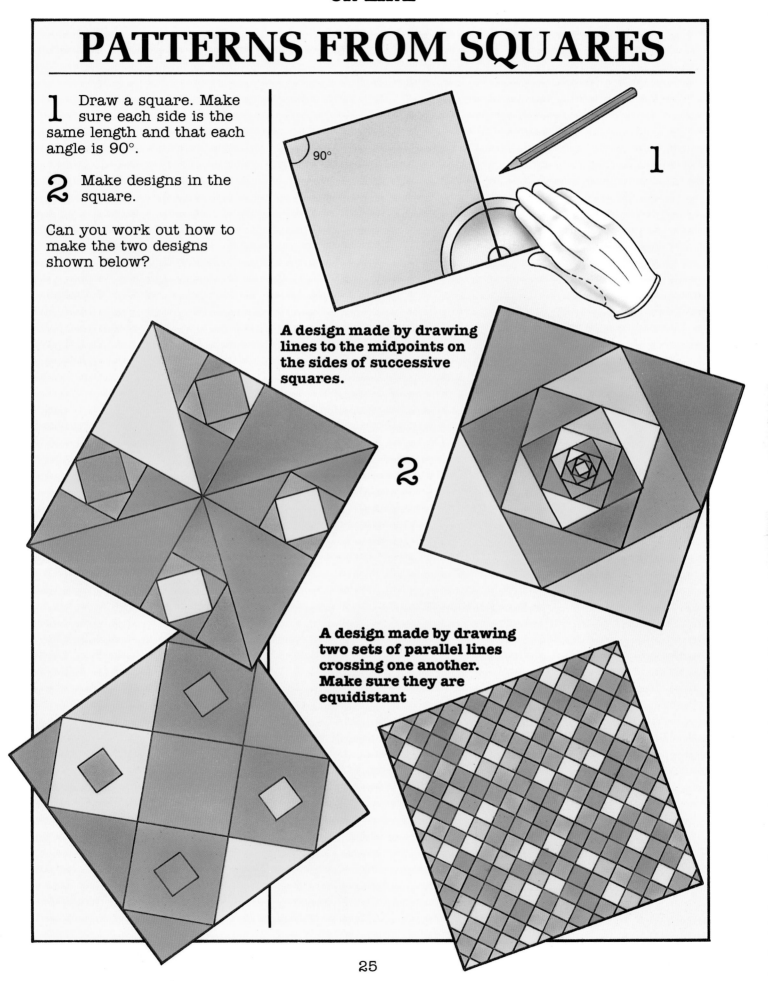

1

A design made by drawing lines to the midpoints on the sides of successive squares.

2

A design made by drawing two sets of parallel lines crossing one another. Make sure they are equidistant

PATTERNS FROM OBLONGS

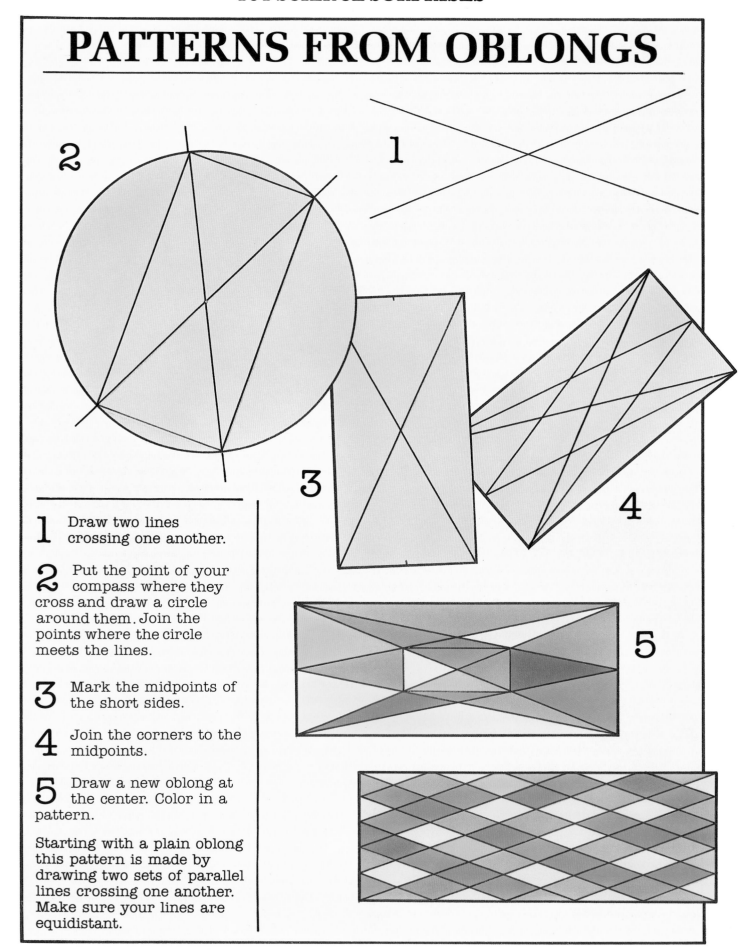

1 Draw two lines crossing one another.

2 Put the point of your compass where they cross and draw a circle around them. Join the points where the circle meets the lines.

3 Mark the midpoints of the short sides.

4 Join the corners to the midpoints.

5 Draw a new oblong at the center. Color in a pattern.

Starting with a plain oblong this pattern is made by drawing two sets of parallel lines crossing one another. Make sure your lines are equidistant.

PATTERNS FROM PENTAGONS

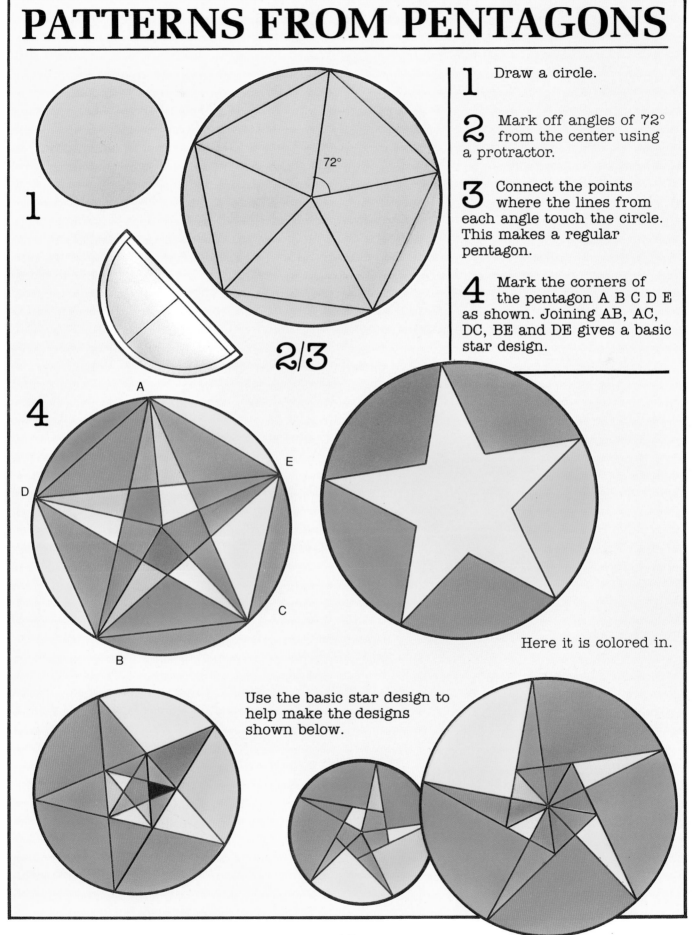

1 Draw a circle.

2 Mark off angles of 72° from the center using a protractor.

3 Connect the points where the lines from each angle touch the circle. This makes a regular pentagon.

4 Mark the corners of the pentagon A B C D E as shown. Joining AB, AC, DC, BE and DE gives a basic star design.

Here it is colored in.

Use the basic star design to help make the designs shown below.

PATTERNS FROM HEXAGONS

1 Draw a circle. Mark a point A on the circle.

2 Place the point of your compass on A and mark point B. Make sure you keep the distance between A and B the same as the radius of your circle.

3 Place the point of your compass on B and mark point C. Again make sure you keep the distance between B and C the same as the radius of your circle.

4 Continue in this way around the circle.

Join the marks to make a regular hexagon.

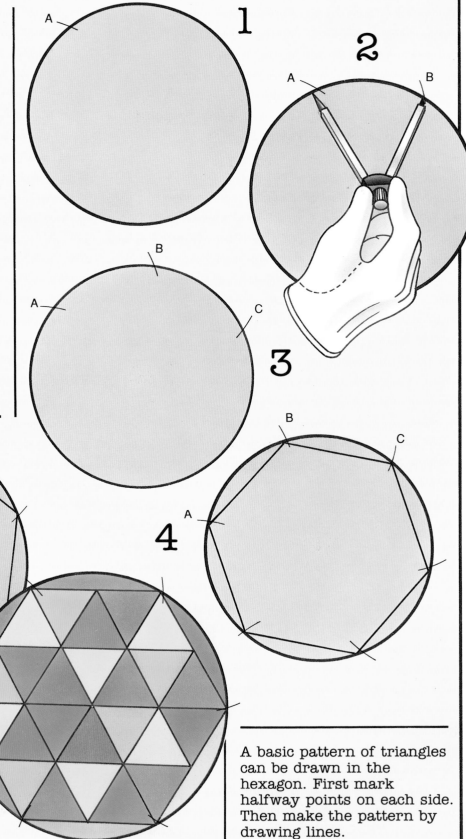

A basic pattern of triangles can be drawn in the hexagon. First mark halfway points on each side. Then make the pattern by drawing lines.

Use the pattern of triangles as a guide to make other patterns.

These are some examples.

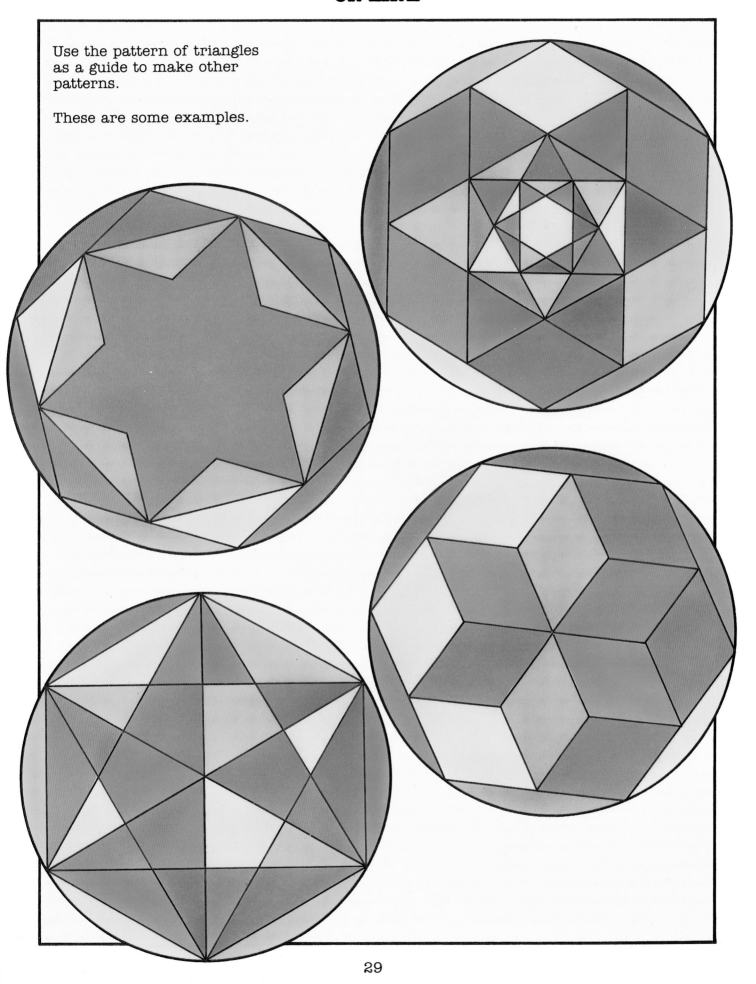

PATTERNS FROM SEPTAGONS

1 Draw a circle.

2 Mark off angles of 51.5° from the center using a protractor.

3 Connect the points where the lines from each angle touch the circle. This makes a septagon.

Can you see how drawing straight lines to various points of the septagon will give you all sorts of patterns? There are some examples shown here.

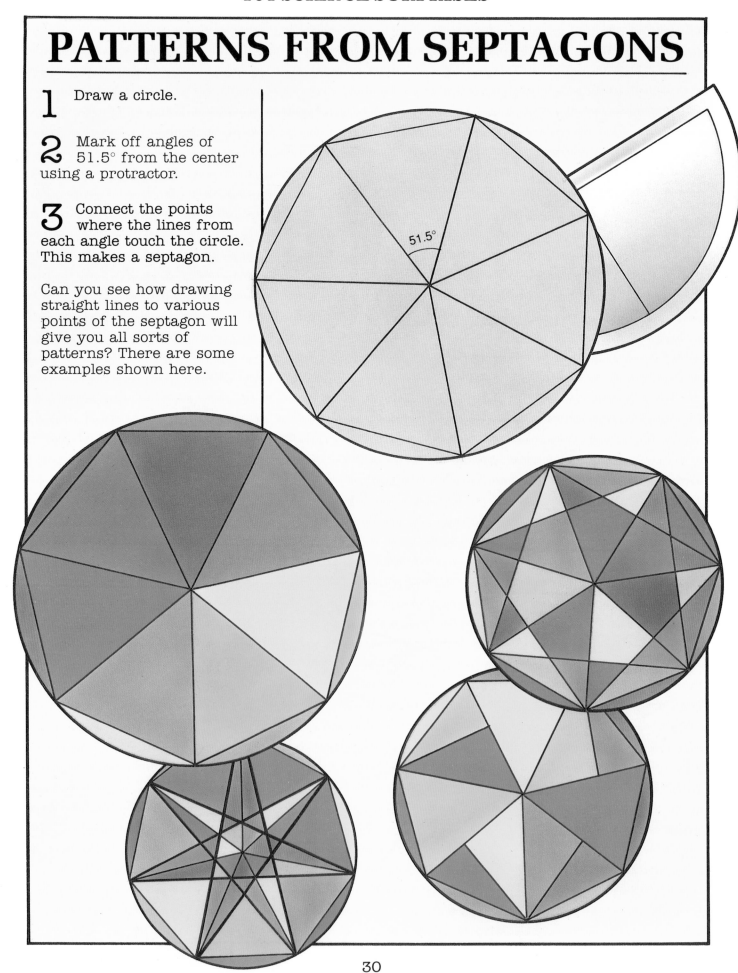

PATTERNS FROM OCTAGONS

1 Draw a circle.

2 Mark off angles of 45° from the center using a protractor.

3 Connect the points where the lines from each angle touch the circle. This makes a regular octagon.

4 Join each point in the octagon to each other point by drawing straight lines. This gives the pattern shown on the right.

Use the pattern to help you make the patterns shown below.

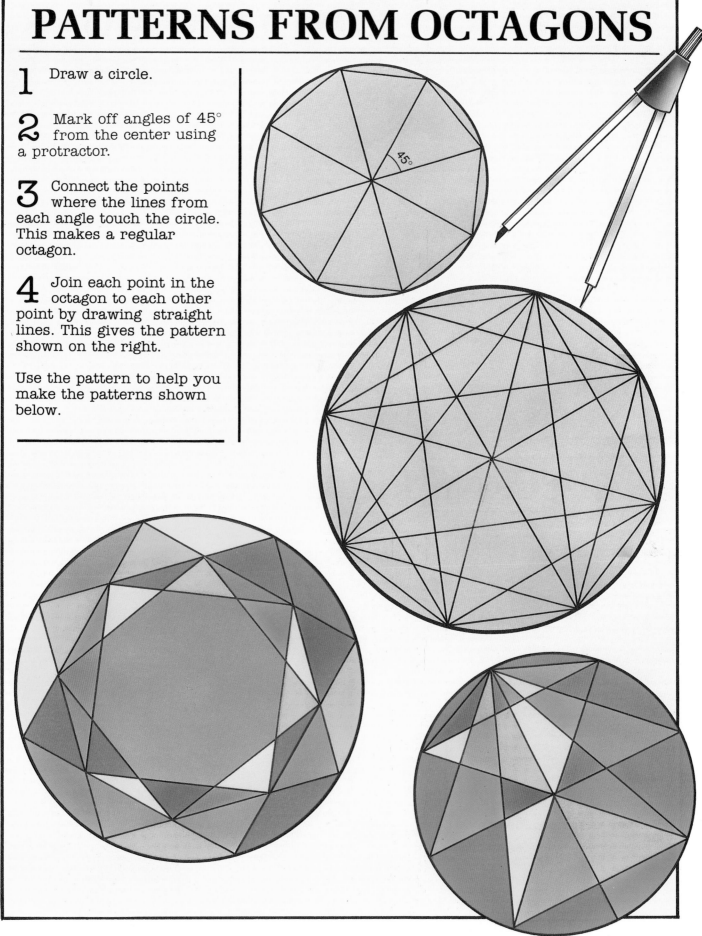

PATTERNS FROM NONAGONS

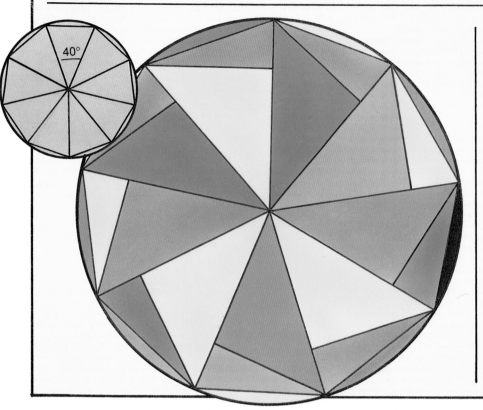

1 Draw a circle.

2 Mark off angles of 40° from the center using a protractor.

3 Connect the points where the lines from each angle touch the circle. This gives a nine-sided figure called a nonagon.

Can you make this design?

PATTERNS FROM DECAGONS

1 Draw a circle.

2 Mark off angles of 36° from the center using a protractor.

3 Connect the points where the lines from each angle touch the circle. This gives a ten-sided figure called a decagon.

4 Using a compass draw another dotted circle inside the first one.

Can you make this design?

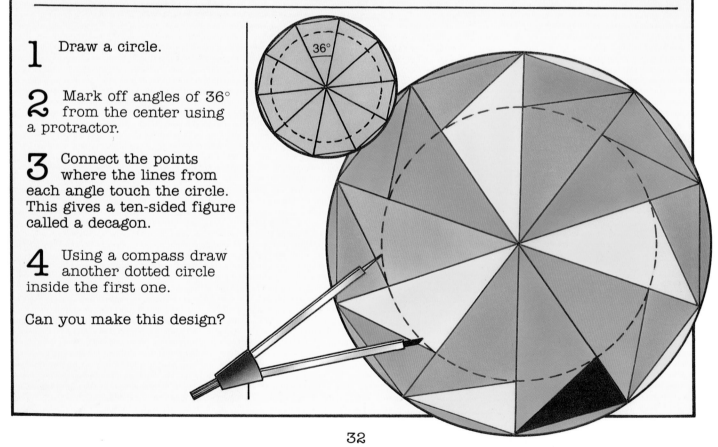

PATTERNS FROM DODECAGONS

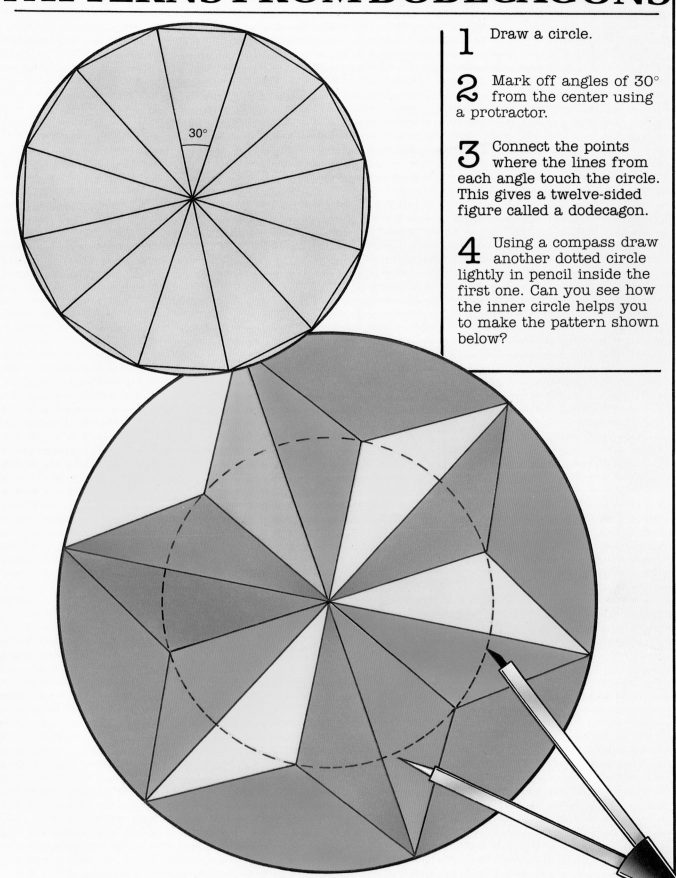

1 Draw a circle.

2 Mark off angles of 30° from the center using a protractor.

3 Connect the points where the lines from each angle touch the circle. This gives a twelve-sided figure called a dodecagon.

4 Using a compass draw another dotted circle lightly in pencil inside the first one. Can you see how the inner circle helps you to make the pattern shown below?

MORE PATTERNS FROM LINES

1 Make an equilateral triangle ABC. (See Page 24.) Make each side 4 in. long.

2 Mark points A_1 B_1 C_1 $1/4$ in. from the corners ABC. Draw the triangle A_1 B_1 C_1.

3 Continue along the sides of the first triangle and mark the points A_2 B_2 C_2 $1/4$ in. from points A_1 B_1 C_1. Draw the triangle A_2 B_2 C_2.

4 Mark the points A_3 B_3 C_3 $1/4$ in. from points A_2 B_2 C_2. Draw the triangle A_3 B_3 C_3.

5 Continue in this way to create a pattern.

Try it with a square.

Color in the patterns.

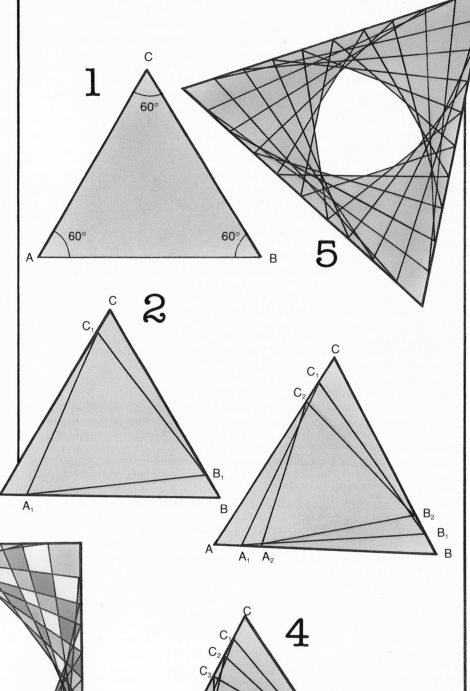

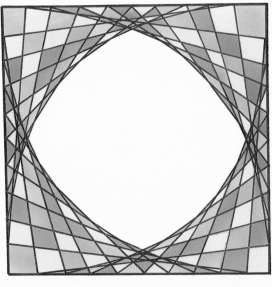

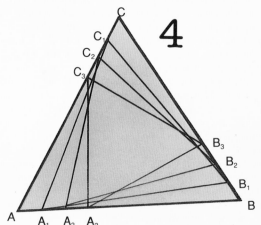

Draw another equilateral triangle with sides 4 in. long. Again, mark points A_1 B_1 C_1 1/4 in. along from ABC, and draw the new triangle. This time mark points A_2 B_2 C_2 1/4 in. from points A_1 B_1 C_1, but move slightly in towards the middle of the first triangle. Do the same with A_3 B_3 C_3, moving a little further in. Carry on in this way. Decorate your patterns.

Do it for the square too.

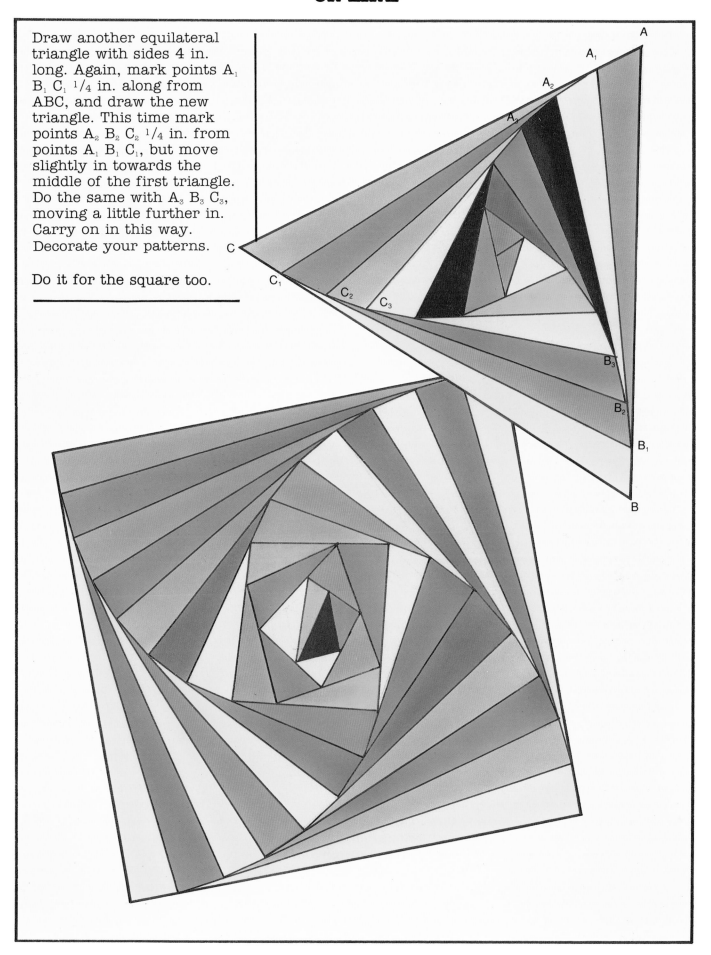

NOTES FOR PARENTS AND TEACHERS

The activities described in this book vary in the degree of skill required by children but all the activities are relatively easy to do, and should be well within the capabilities of children in the 7–13 age range. There is a strong scientific or mathematical basis to each of the activities. For children the fun of making and doing is all. For parents and teachers the following notes may be of interest in making the most of what lies beyond all these things to do.

The activities all involve the following of a line. The patterns created in the latter part of the book are all obtained by drawing straight lines.

Page 7 Snails move on a slime trail. It helps reduce friction as they move and it helps them to cling to things. Snails are more active at night.

Pages 8–9 Labyrinths are known from ancient Greek times. More recently patterns were made on the floors of cathedrals such as Chartres where penitents would follow a line on the floor, reciting their prayers at fixed stations. From the churches they moved out into gardens that provided private walks flanked by hedging. Complex mathematical patterns have been worked out for such walks.

Pages 10–11 Here are two mazes that demand some simple technological skills. One tests dexterity of hand, the other relies on the sense of touch.

Page 12 This activity takes children into forming lines of figures. The figures illustrate what is mathematically called "translational symmetry," It is the sort of symmetry you get whenever you have a repeated pattern, for example, in the bars of railings, and in the segments of a caterpillar.

Page 13 The letter balance depends for its action on the elasticity of a rubber band. The stretch of the band will vary with the weight hung at its end.

Page 14 A tightly drawn piece of nylon fishing line will give a note when plucked. The tighter the line the higher the note.

Page 15 Sound travels through gases, liquids and solids. Here it is channelled along a solid-thin string. It is important to have the string taut when the telephone is in action so that the vibrations can travel easily along the string. Tell children there is no need to shout into the tin-can, the idea is to speak quietly and distinctly and to let the string and the can do the work. The thinner the tin-cans the better.

Page 16 The "trick" in getting the Cartesian diver to work is to make sure it is only just floating.

Squeezing the bottle then forces water into the diver making it too heavy to float, and it thus sinks. Releasing the bottle in turn releases water from the diver. It is now light enough to float again and it bobs up in the water.

Page 17 This activity makes use of a technique used by chemists. It is called chromatography. As the water reaches the red ink from the felt-tip pen it causes the pigments in the ink to dissolve and be carried away on the water front. Some of the pigments are made up of smaller particles than others. These dissolve first. The pigments thus separate from one another. The different colors making up the red ink separate out leading to the movement along the strip of paper. Chemists would use such a technique to identify the pigments making up any particular ink.

Pages 18–19 Both these toys depend on the energy stored in elastic to cause movement. In one case it fires a coin, in the other it kicks a soccer ball.

Pages 20–22 Weaving must be one of the oldest of technologies. It is used it to make garments and roofing material.

Page 23 A.F. Mobius was a Swedish geometer (1790–1860). He introduced the Mobius strip, a band with one side and one edge. Test your loop with a half twist in it to verify this. As you follow the edge you will see that it is a deformation of a single circle. Cutting (as you are asked to do in the activity) along the central line enables you to obtain the circle. Fascinating!

Pages 24–35 These pages yield a wealth of patterns that can be obtained merely by drawing straight lines. They take children through a number of geometrical constructions and intrigue them with the enormous variety of patterns that mathematical pursuit often yields.

2
on Nature

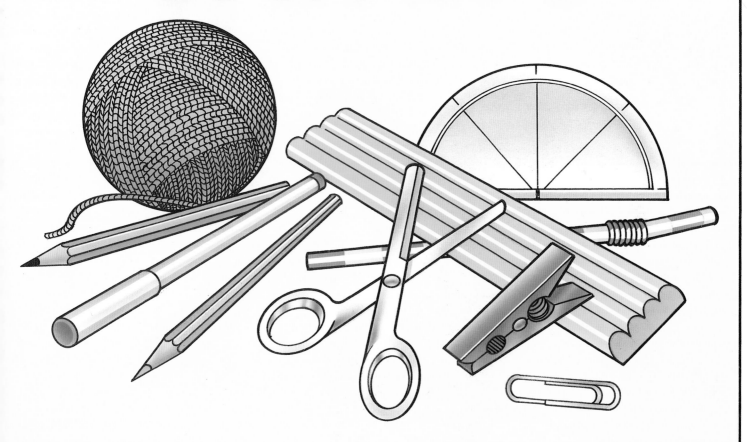

INTRODUCTION

All the fun things to make and do in this section are
related to the natural world. You can make a family
portrait gallery, have a go at playing detective, and create
models that show what is inside our bodies.

It's fun to make gifts for people, and you'll find some
ideas for a seashell picture, a lacquered shell box, an
animal dangler for a young child's bedroom and a
miniature garden. Then there are snowflake patterns to
put on a windowpane, pressed flower pictures for the wall,
a bird feeder for the garden, moving toys, leaf prints, and
much more.

Everything you will need to make these things is
listed on page 101.

SHELL WORK

Seashells can be used to make delightful pictures and to decorate boxes and pots.

A SEASHELL PICTURE

1 Select the shells you are going to use and arrange them on a large piece of paper until you have a design you like.

2 Cut a piece of cardboard or mounting board the right size for your picture.

3 Put a little white glue along the outside edge of the small shells and stick them directly to your board. With a large shell you need to make a wad of tissue paper and glue it inside the shell. You then apply glue to the wad and stick it to the board.

tissue pad

SEASHELL BOXES AND POTS

1 Select a variety of suitable shells.

2 Mix some Polyfilla in an old margarine tub. Put a little of the powder in the tub and add a small amount of water, stirring constantly.

3 When you have a thick paste smear it thickly over your box or pot.

4 Work quickly, gently pushing the shells into the paste while it is still wet.

5 Leave to dry.

6 Paint the shells with clear varnish to give your finished product a sheen.

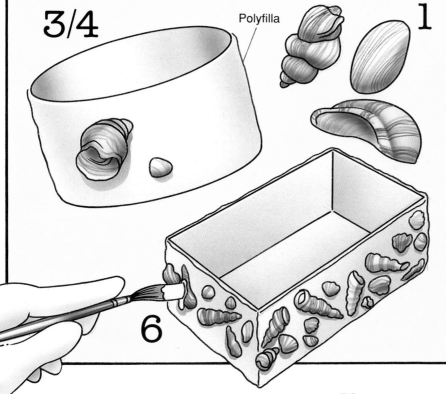

Polyfilla

3/4

1

6

FOOD CHAIN DANGLERS

Some animals eat other animals. These in turn may eat other animals or they may eat plants. This makes a food chain.

This dangler shows a fox. It has eaten a rabbit, which has eaten a lettuce.

You can trace this dangler.

1 Trace the fox first on to oaktag. Cut out his 'belly'.

2 Now trace the rabbit on to oaktag. Again cut out the 'belly'.

3 Trace the lettuce on to oaktag and cut it out.

4 Dangle the lettuce inside the rabbit with thin thread. Then dangle the rabbit and the lettuce inside the fox.

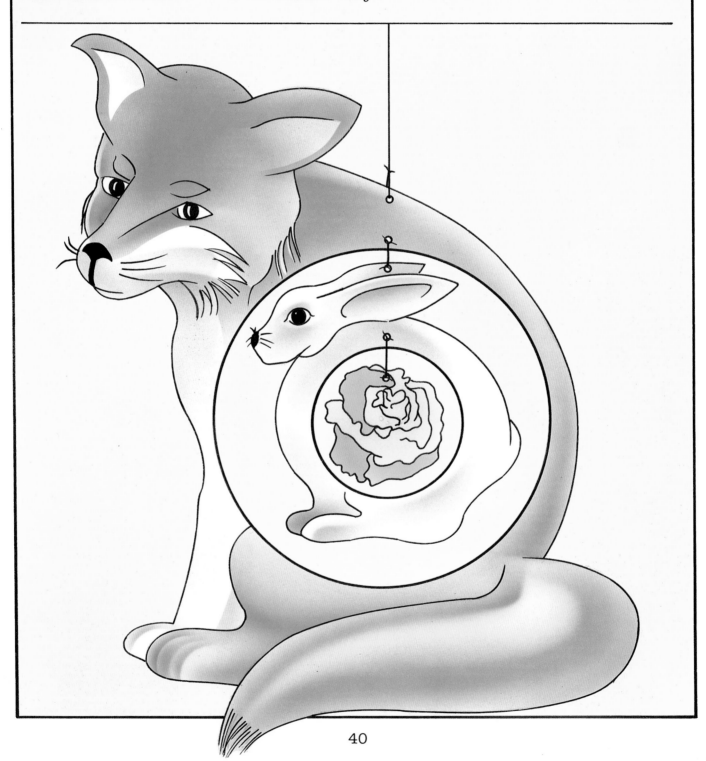

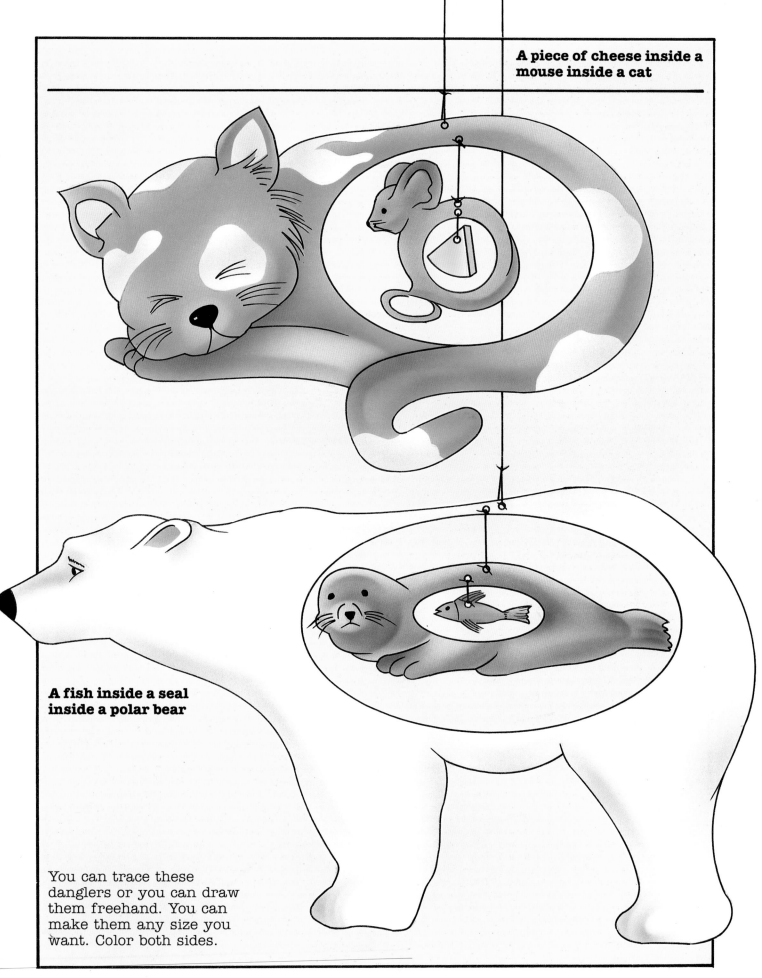

A piece of cheese inside a mouse inside a cat

A fish inside a seal inside a polar bear

You can trace these danglers or you can draw them freehand. You can make them any size you want. Color both sides.

41

A LIFE-CYCLE WHEEL

There are four stages in the life history of a butterfly. They are egg, caterpillar, chrysalis and imago (adult).

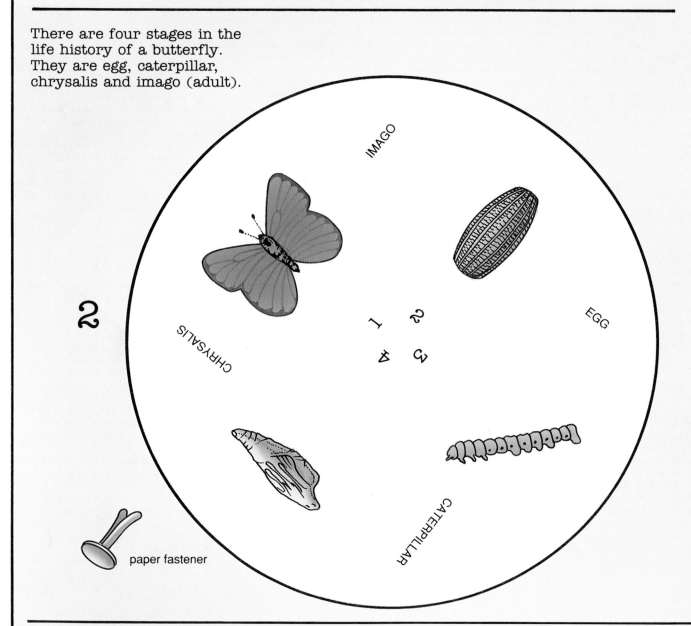

IMAGO

EGG

CHRYSALIS

CATERPILLAR

2

1 2
4 3

paper fastener

1 Cut two pieces of oaktag the same size as the discs shown here.

2 Trace the stages in the life history of the butterfly shown above on to one piece of oaktag. Trace them in position exactly as shown. Trace the names and numbers too.

3 Trace the shaded portions shown on the large circular diagram opposite on to the second piece of card. Cut them out.

4 Put this piece of card on top of the one showing the stages in the life-cycle.

5 Fasten the two cards at the center with a paper fastener. You can decorate the top disc.

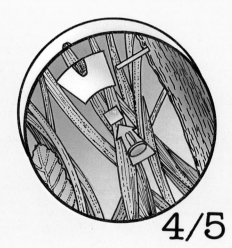

4/5

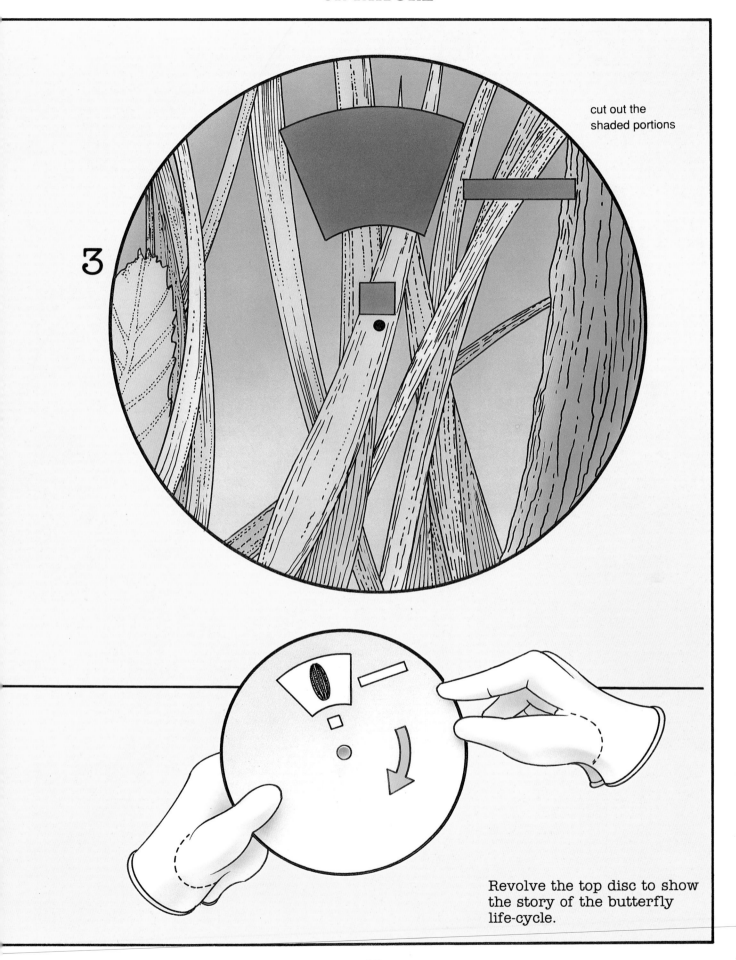

3

cut out the
shaded portions

Revolve the top disc to show
the story of the butterfly
life-cycle.

FAMILY PORTRAITS

Did you know that before the age of photography it was fashionable to make and mount silhouette portraits? Skilled portrait makers did it freehand by cutting a portrait from a single piece of paper with scissors. This is difficult. Here is a simple method to do it using a lamp.

1 Fix a sheet of drawing paper to the wall. You can hold it in place with masking tape.

2 Sit a member of the family or a friend comfortably in front of the sheet with the head sideways onto your sheet.

3 Darken the room and use a lamp to cast a shadow of the head onto the sheet. You will need to move the lamp back and forth to find the best position to give you a sharp shadow outline of the head.

4 Ask the sitter to keep as still as possible. Draw the silhouette.

5 You can paint or crayon the silhouette. If you use paper that is colored on one side and white on the other, you can turn it round to give a colored silhouette when you cut it out.

Make a family portrait gallery.

A TOUCH TESTER

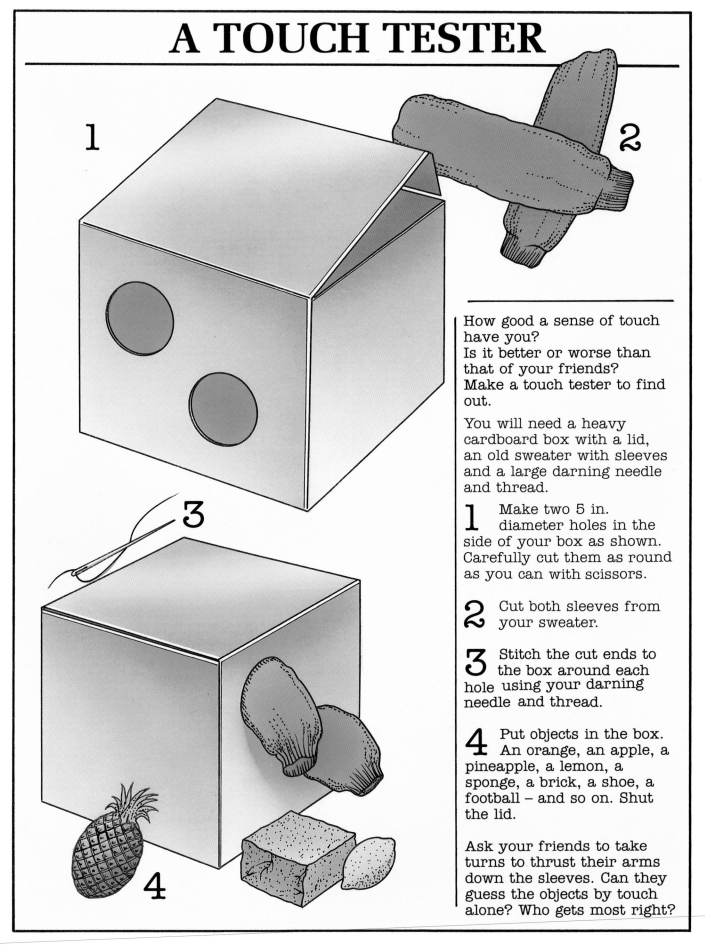

How good a sense of touch have you?
Is it better or worse than that of your friends?
Make a touch tester to find out.

You will need a heavy cardboard box with a lid, an old sweater with sleeves and a large darning needle and thread.

1 Make two 5 in. diameter holes in the side of your box as shown. Carefully cut them as round as you can with scissors.

2 Cut both sleeves from your sweater.

3 Stitch the cut ends to the box around each hole using your darning needle and thread.

4 Put objects in the box. An orange, an apple, a pineapple, a lemon, a sponge, a brick, a shoe, a football – and so on. Shut the lid.

Ask your friends to take turns to thrust their arms down the sleeves. Can they guess the objects by touch alone? Who gets most right?

DETECTIVE WORK

Any burglars around your house? Check outside for footprints.

1 Put a strip of cardboard around your footprint. Hold the ends together with a paper clip.

2 Mix some plaster of Paris in an old bowl or other container. A large throw-away food tray will do. Add the plaster powder to the water. Stir all the time until the mixture is the consistency of thick cream.

3 Once it is ready do not delay. Pour it straight into the cardboard mold placed around your footprint, filling it to the top.

4 Leave it overnight and then strip away the card. Turn the cast over to see the print underneath. Clean your hardened cast with an old toothbrush.

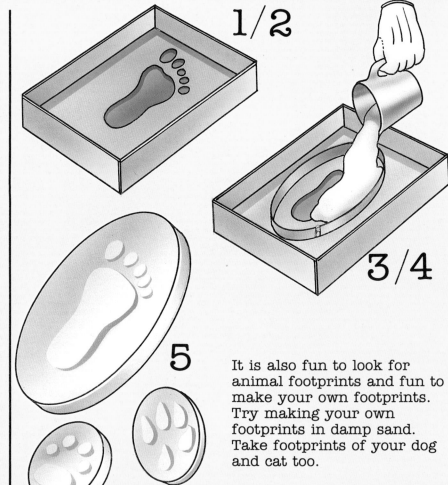

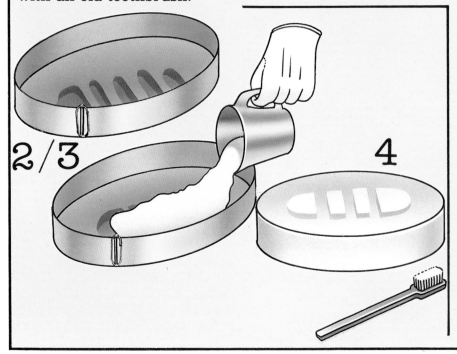

It is also fun to look for animal footprints and fun to make your own footprints. Try making your own footprints in damp sand. Take footprints of your dog and cat too.

1 Start with a tray of firm damp sand.

2 Make a footprint.

3 Press a strip of cardboard in the sand.

4 Pour in the plaster of paris.

5 Clean the dry cast with an old toothbrush. Paint the cast. Choose a different color for the background.

Compare your animal casts with any other prints in your area. Do they belong to your pets or do they belong to intruders?

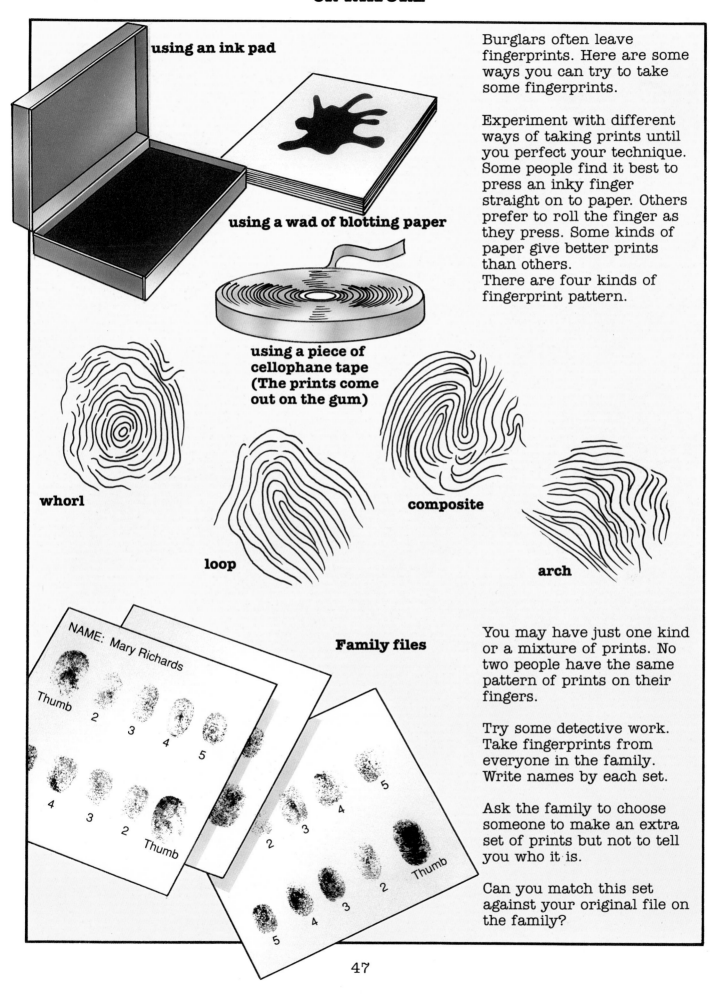

using an ink pad

using a wad of blotting paper

using a piece of cellophane tape (The prints come out on the gum)

whorl

loop

composite

arch

Burglars often leave fingerprints. Here are some ways you can try to take some fingerprints.

Experiment with different ways of taking prints until you perfect your technique. Some people find it best to press an inky finger straight on to paper. Others prefer to roll the finger as they press. Some kinds of paper give better prints than others.
There are four kinds of fingerprint pattern.

Family files

NAME: Mary Richards

Thumb 2 3 4 5

4 3 2 Thumb

5 4 3 2 Thumb

5 4 3 2

You may have just one kind or a mixture of prints. No two people have the same pattern of prints on their fingers.

Try some detective work. Take fingerprints from everyone in the family. Write names by each set.

Ask the family to choose someone to make an extra set of prints but not to tell you who it is.

Can you match this set against your original file on the family?

47

SEED PICTURES

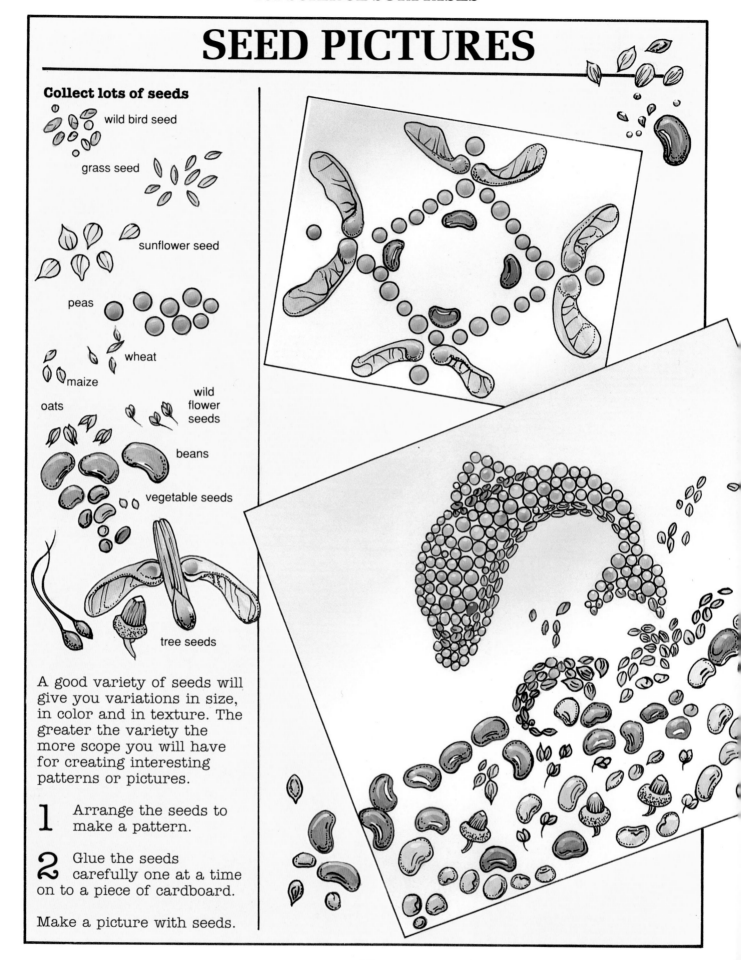

Collect lots of seeds

wild bird seed

grass seed

sunflower seed

peas

wheat

maize

oats

wild flower seeds

beans

vegetable seeds

tree seeds

A good variety of seeds will give you variations in size, in color and in texture. The greater the variety the more scope you will have for creating interesting patterns or pictures.

1 Arrange the seeds to make a pattern.

2 Glue the seeds carefully one at a time on to a piece of cardboard.

Make a picture with seeds.

SEEDS TO GROW AND EAT

Mustard and cress are the seeds that you need. Save two eggshells from breakfast.

1 Use felt-tip pens to paint a face on each shell.

2 Put some moist potting soil in each shell.

3 Plant mustard seeds in one and cress seeds the other. Keep them moist but do not overwater.

They will grow fine heads of mustard and cress "hair" within a few days.

Try it in other containers.

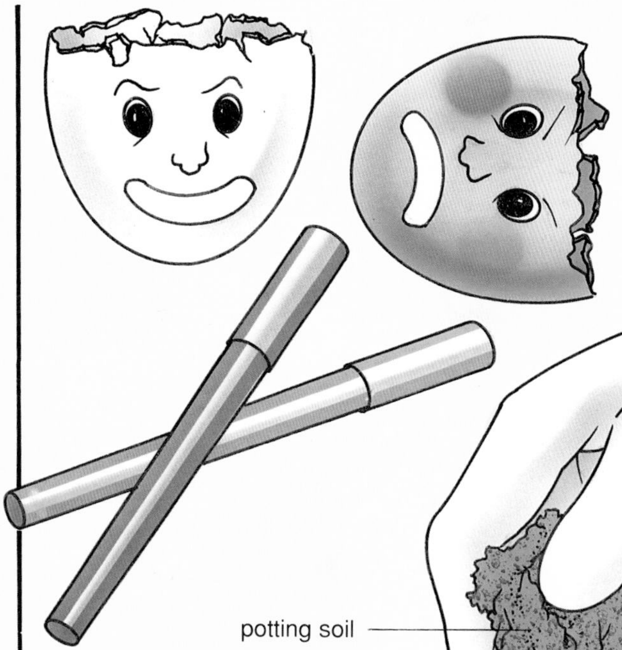

1

potting soil

2

3

An orange

A Plasticine hedgehog

A MINIATURE GARDEN

1 Plan your garden in a large metal tray. Mark the outline of the plots and the paths with chalk or felt-tip pen.

2 Use some ready mixed sand and cement to make paving slabs. Put the mixture in an old bowl and add water slowly, stirring all the time. Once the mixture is the consistency of thick oatmeal pour it into the matchbox trays. These act as molds. When the mixture is set tip the slabs out.

3 Make Plasticine edges for the paths.

4 Fill the paths between the Plasticine edges with sand. Set the paving stones into the sand.

5 Plant one of the areas with grass seed and another area with mustard or cress. Fill these two areas with potting compost. Pat it down. Plant with seeds. Plant the cress in rows.

6 Fill the area around the pond with gravel. Fill the remaining area with miniature pots of cacti or small herbaceous plants. You can embed them with some pebbles in a mixture of sand and cement if you want.

7 As the grass grows, you can trim it with scissors. Fill the pond with water.

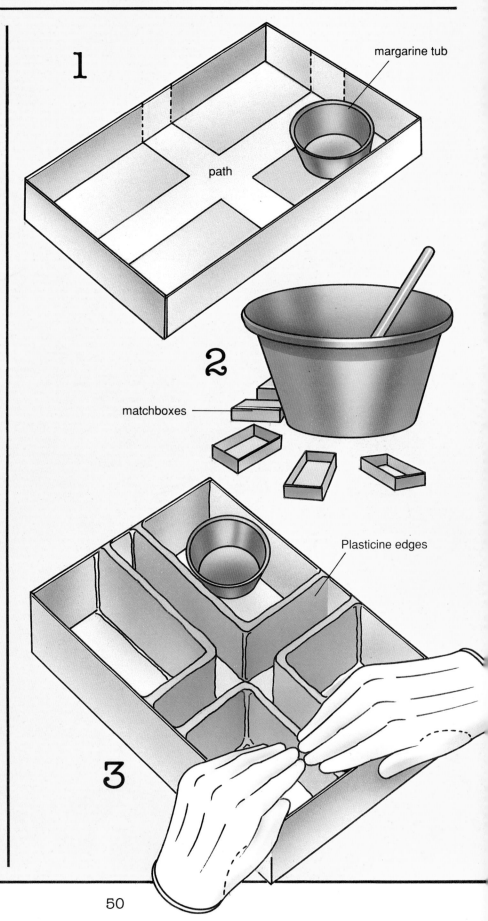

margarine tub

path

matchboxes

Plasticine edges

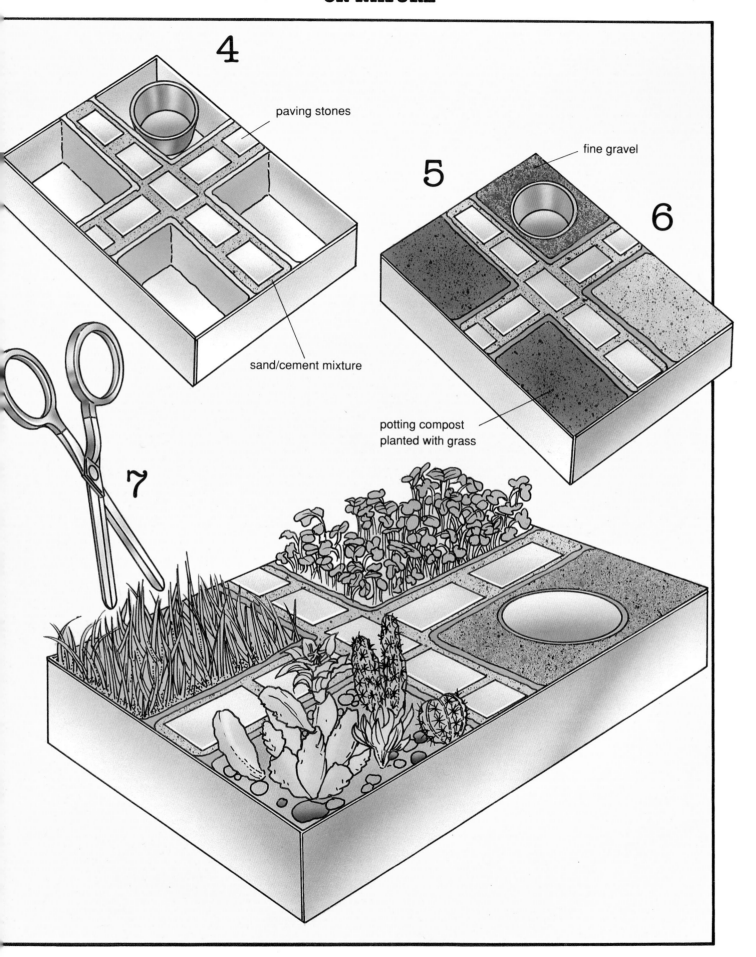

4

paving stones

sand/cement mixture

5

fine gravel

6

potting compost
planted with grass

7

BOTTLE GARDENS

Use a large kitchen spoon and fork to put plants into the jars to make your bottle gardens.

A jar garden

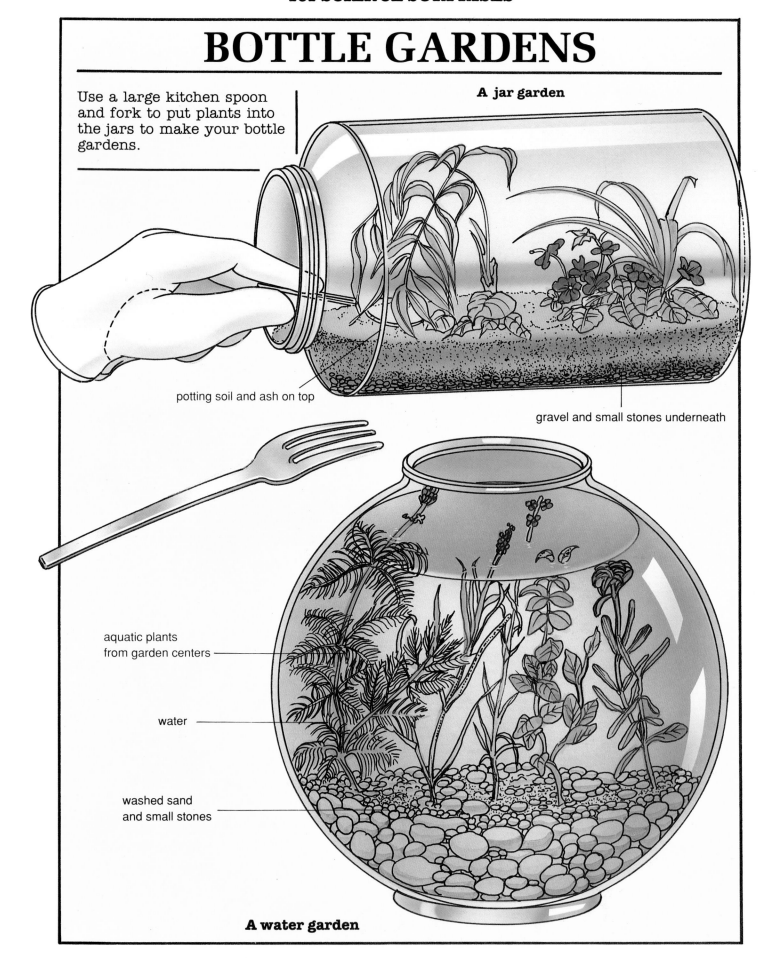

potting soil and ash on top

gravel and small stones underneath

aquatic plants from garden centers

water

washed sand and small stones

A water garden

MUSHROOM PRINTS

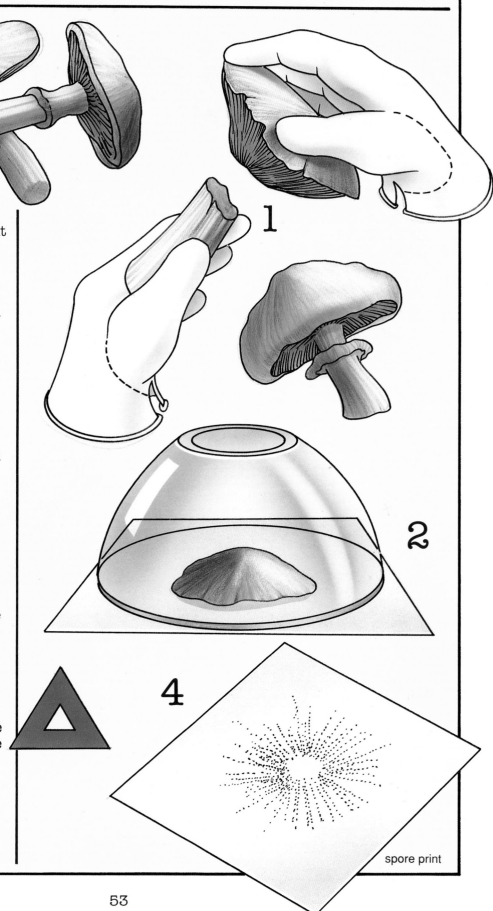

Autumn is the best season for collecting mushrooms and toadstools. Ask an adult which mushrooms are safe to pick. **Don't taste any.**

1 Cut the stalk off each one, just below the cap.

2 Put the cap, gills downward, on a piece of paper. Cover it with a clear glass basin to keep out drafts.

3 Leave it overnight. Lift the cap gently and carefully the next day. You will find that spores have fallen from the gills to make a pattern.

4 Spore color varies with the kind of mushroom or toadstool you have got. You may have white or pink spores. These show up best on black paper. Black, purple or brown spores are best shown on white paper.

5 If you want to make permanent records use waxed paper. Once you have your spore print you can warm the paper above a radiator. Do it gently and take care. The wax in the paper will melt, when it cools again it hardens and traps the spores.

spore print

SNOWFLAKES

Did you know that the famous scientist Robert Hooke first made illustrations of snowflakes seen through a miscroscope in 1665?

Snow crystals have a hexagonal pattern.

1 Draw a circle on card and mark off 60° angles using a protractor.

2 Draw in a pattern.

3 Cut out the pattern and hang it by a thread.

Try other designs.

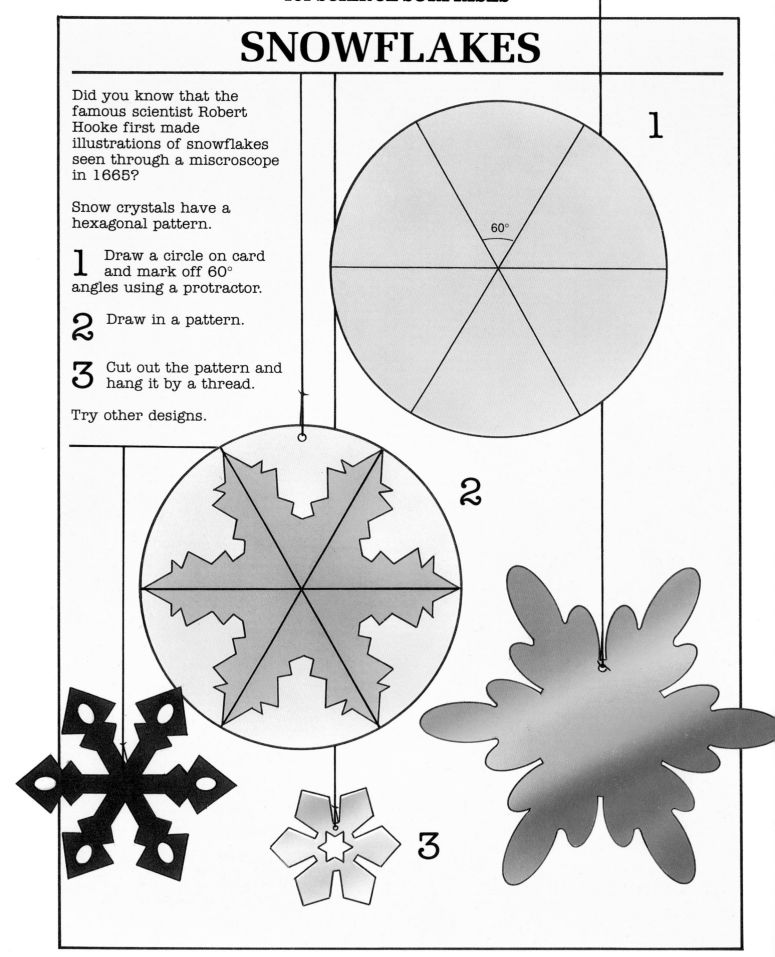

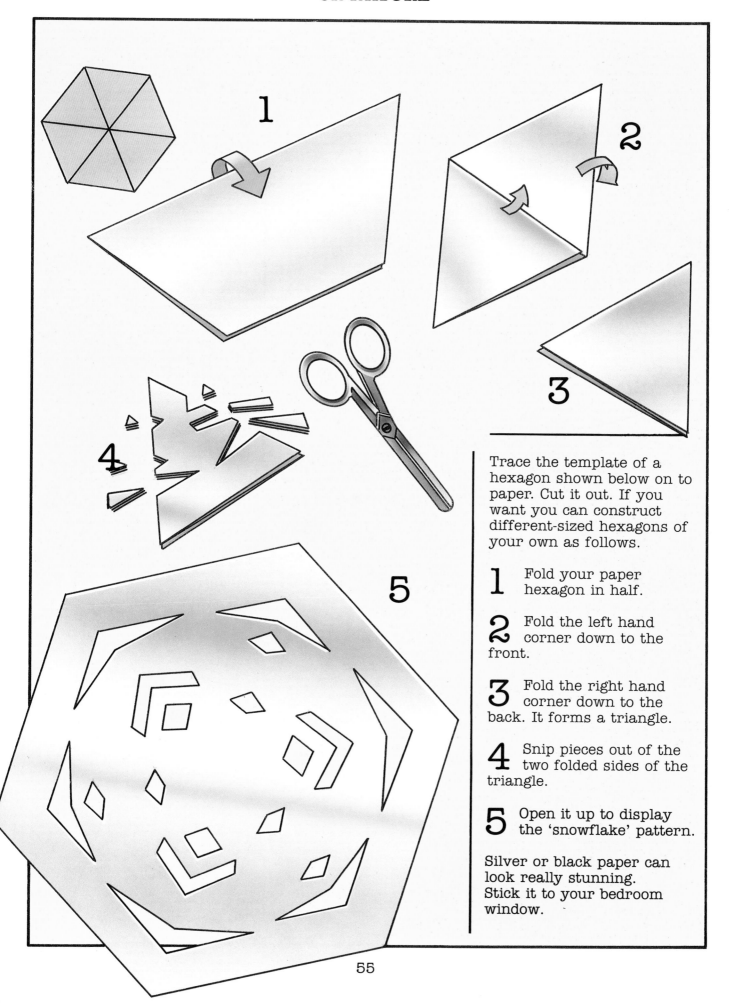

Trace the template of a hexagon shown below on to paper. Cut it out. If you want you can construct different-sized hexagons of your own as follows.

1 Fold your paper hexagon in half.

2 Fold the left hand corner down to the front.

3 Fold the right hand corner down to the back. It forms a triangle.

4 Snip pieces out of the two folded sides of the triangle.

5 Open it up to display the 'snowflake' pattern.

Silver or black paper can look really stunning.
Stick it to your bedroom window.

MAKING LEAF PRINTS

Prints of leaves make effective records of some of the most interesting forms in nature. Recently fallen autumn leaves are best to use because they dry and harden quickly. You can make prints of leaves all through the year. It is easier to do it with simple leaves rather than divided ones.

SCRIBBLE PRINTS

1 Put the leaf on a newspaper with the bottom surface up.

2 Cover it with a sheet of white paper.

3 Hold the paper firmly so that it does not slip. Rub with a crayon, making sure that you do edges, stalks and veins well.

4 Cut out the print and mount it on colored paper.

SHOE POLISH PRINTS

Red, brown and tan shoe polishes can match the color in autumn leaves.

1 Smear the backs of the leaf with polish using a cotton ball or a finger. Spread evenly and sparingly.

2 Place the leaf, polish side down, on a sheet of white paper. Cover it with another sheet of paper.

3 Rub gently on the top piece of paper. Lift the leaf to reveal a colored print. Cut it out.

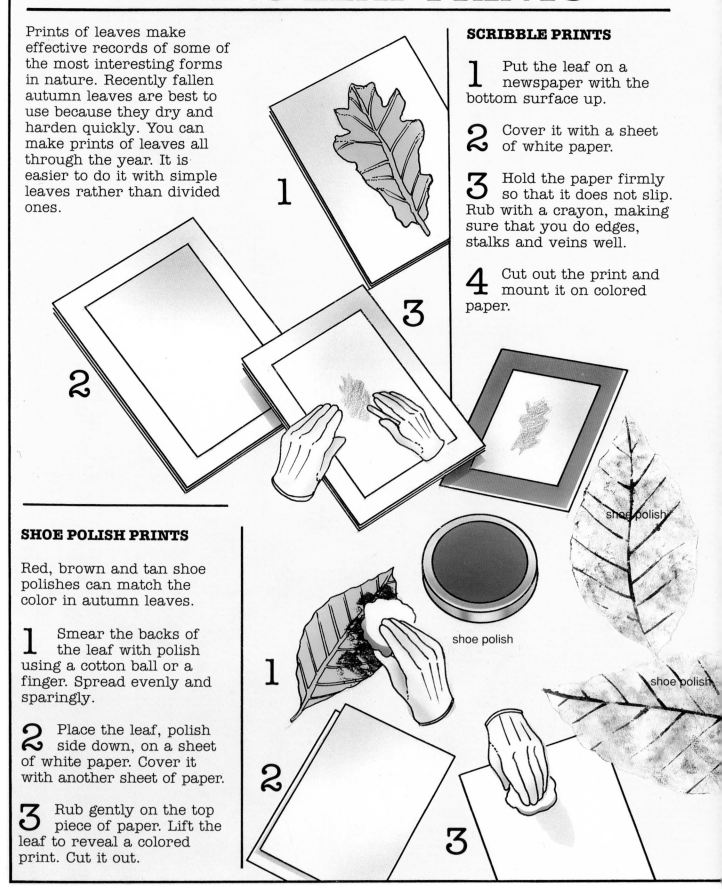

shoe polish

56

SPATTER PRINTS

This technique gives very attractive leaf outlines.

1 Place a leaf on to a sheet of white paper using plenty of newspaper to protect the surface you are working on.

2 Run the blade of a blunt knife over the bristles of a paint-covered toothbrush to flick the wet paint.

3 Remove the leaf to reveal the white silhouette.

4

PRESSED FLOWER PICTURES

Flowers and leaves can be pressed to make pictures.

1 Gather your material carefully and place each piece well spaced from others between two sheets of blotting paper.

2 Make sure the material lies flat.

3 Put the blotting paper sheets with the drying material between the pages of a large old book.

4 Weight the top with bricks and leave for a month.

HINTS

With large flowers such as roses it is best to separate each petal and press singly.

With delicately colored petals lay one on top of the other. When they dry, you will get a deeper color.

With flowers like daisies press the whole flower head.

With flowers like marigolds, which have a thick center, flatten the center with the back of a spoon before you press.

bricks

rose petals

PICTURE MAKING

1 Cut two pieces of heavy paper to the same size. One to arrange your picture on and one to stick the pieces to.

2 Use a paint brush to move the pieces about but handle them as little as possible.

3 When the arrangement is ready stick each piece with a tiny dab of glue. Stalks will need several dabs along their length.

JUMPING FROGS

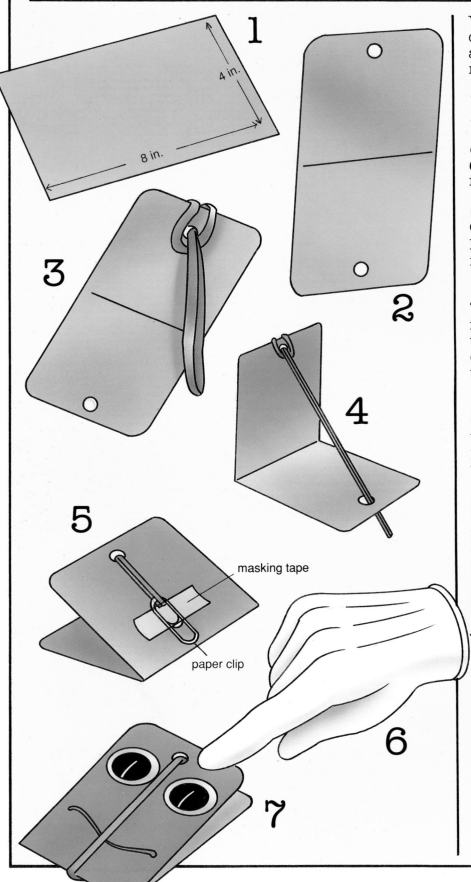

masking tape

paper clip

You will need some cardboard, a rubber band, a paper clip and some masking tape.

1 Cut the cardboard to size.

2 Round off the corners. Mark the center and make a hole at each end.

3 Push the rubber band through one of the holes and loop it through itself to pull it tight.

4 Bend the cardboard backwards and forwards a few times in the middle. Pull the loose end of the rubber band through the other hole.

5 Attach the loose end to a paper clip. Stretch the elastic band and fix the paper clip to the card with the masking tape.

6 Turn your "frog" inside out so that the paper clip is on the inside and the rubber band stretched over the outside. Put it flat on the table with your finger holding it down. Let go and it will jump in the air.

7 Decorate your frog.

PECKING BIRDS

1 Trace the templates below on to thick cardboard.

2 Fix the pieces together with paper fasteners.

3 Hold one strip in your right hand and the other in your left hand. Move the strips gently back and forth to make the birds peck.

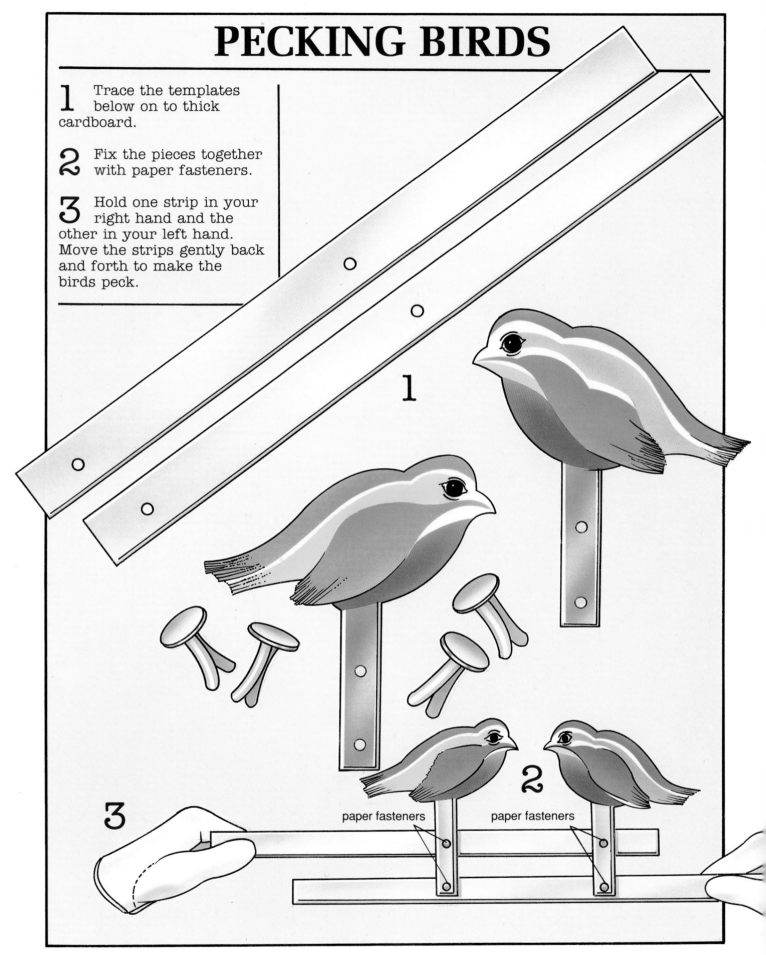

paper fasteners   paper fasteners

A DRUMMING WOODPECKER

This woodpecker pecking its way down a pole is a wonderful toy, and it's easy to make!

1 Twist a 10 in. length of galvanized wire around a 1 yd. length of $3/8$ in. thick dowel to form a spring. About $1^1/_2$ in. from the top end, twist and bend the wire up at an angle.

2 Try the spring on the rod. It should slowly bounce its way down once you let it go. If it's wound too tightly it will stick. If it's wound too loosely it will drop too quickly.

3 Make a large woodpecker out of colored Plasticine. Push it on the end of the spring and set it at the top of the dowel. Watch it peck its way down.

If the woodpecker is too slow, loosen the wire coil. If it's too fast, tighten the wire coil. Keep adjusting it until you get the speed just right.

Plasticine

3

$3/8$ in. thick dowel

galvanized wire

1

2

A BIRD FEEDER

1 Carefully remove the top of a detergent bottle and cut off the stopper.

2 Carefully cut four slits, each $3/16$ in. wide, down the bottle as shown. You can use scissors.

3 Pass a piece of string through the top. Make two or three good knots to hold it in place.

4 Cut small slits in the bottle near the base in which to insert the pencils. These will make a perch for the birds.

5 Fill the bottle with peanuts. Replace the top and hang the feeder in a suitable place for the birds to feed. Make sure it is placed away from cats and other predators.

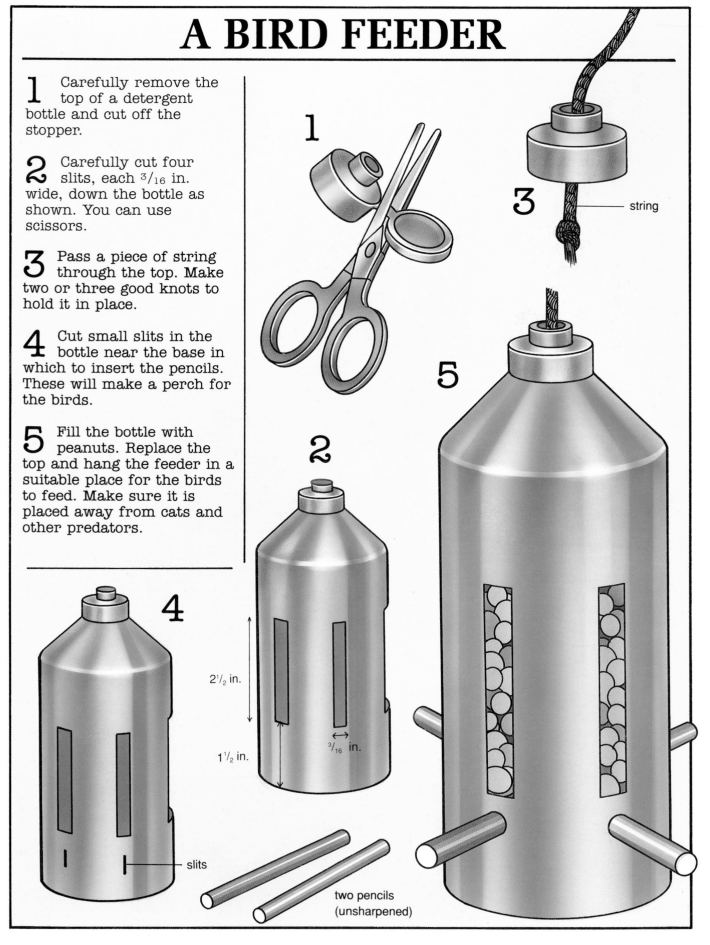

string

$2^{1}/_{2}$ in.

$3/16$ in.

$1^{1}/_{2}$ in.

slits

two pencils (unsharpened)

A QUILL PEN

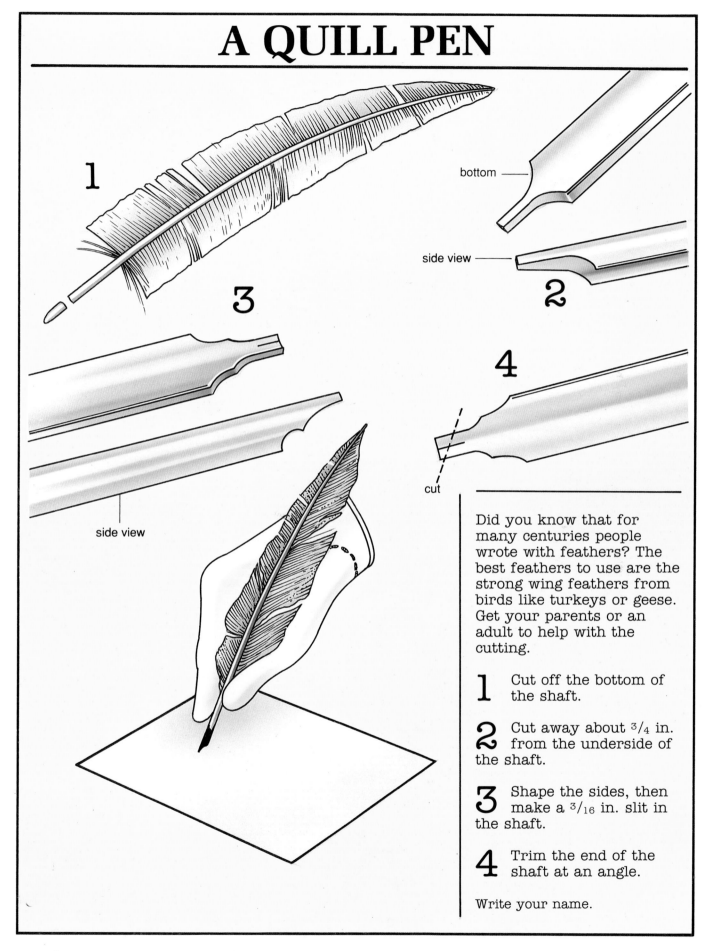

1

bottom

side view

2

3

side view

4

cut

Did you know that for many centuries people wrote with feathers? The best feathers to use are the strong wing feathers from birds like turkeys or geese. Get your parents or an adult to help with the cutting.

1 Cut off the bottom of the shaft.

2 Cut away about 3/4 in. from the underside of the shaft.

3 Shape the sides, then make a 3/16 in. slit in the shaft.

4 Trim the end of the shaft at an angle.

Write your name.

A SKELETON MODEL

skull

hip girdle

shoulder blades

ribs

straw

matchsticks

1 Trace the "skull", "ribs" and "hip bones" on to paper. Cut them out.

2 Bend them into cylinders and glue the edges together so that they become solid models.

3 Flatten each cylinder a little. Use matchsticks for the limbs and collar bones and a flattened drinking straw for the backbone.

Glue the model on to a colored piece of cardboard in this order:

a glue rib cage, hip bones and backbone in position;

b cut out the shoulder blades and glue them behind the rib cage;

c glue the collar bones in position;

d fix the skull at the top of the backbone;

e finish by sticking down the arm and leg bones.

d

c

b

a

e

64

AN INSIDE US MODEL

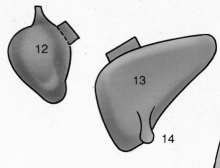

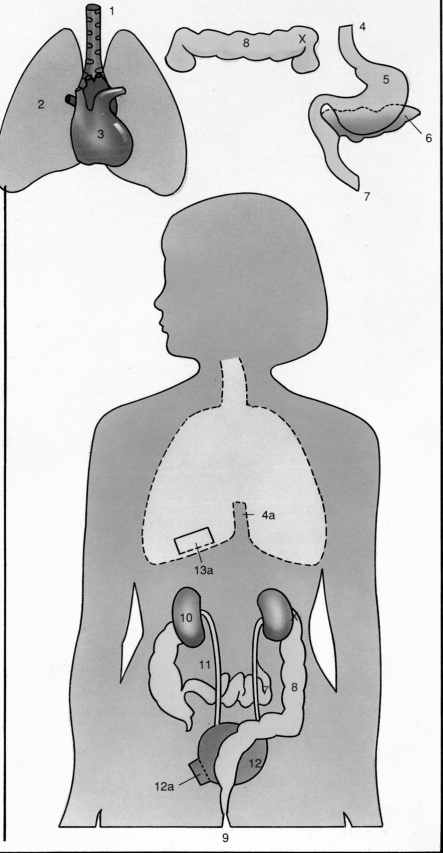

Key

1. Windpipe 2. Lungs 3. Heart
4. Throat 5. Stomach 6. Pancreas
7. Small intestine 8. Large intestine
9. Anus 10. Kidney 11. Tube from kidney
to the bladder 12. Bladder 13. Liver
14. Gall bladder

1 Trace the outline of the body with its internal organs as shown opposite.

2 Trace the separate organs shown above.

3 Color the organs as follows:
  Yellow – the digestive tract i.e. throat, stomach, small intestine, large intestine
  Green – liver, gall bladder, pancreas
  Brown – kidney, bladder
  Pink – windpipe, lungs
  Red – heart.

4 Cut out the body-outline. Cut out the separate organs.

5 Paste the organs down like this – tip of foodpipe (4) on 4a – tip of piece of large intestine marked X on part X – tab of bladder (12) on 12a – tab of liver (13) on 13a – stick the lungs and heart between the dotted lines.

MAGNIFIERS

Biologists often use a hand-lens or a microscope to look at small things.

Here are three magnifiers to make.

A The first one is simply made by bending a wire loop. Put a drop of water or cooking oil on the loop.
Hold it above some newsprint.
Letters seen through the loop are magnified.

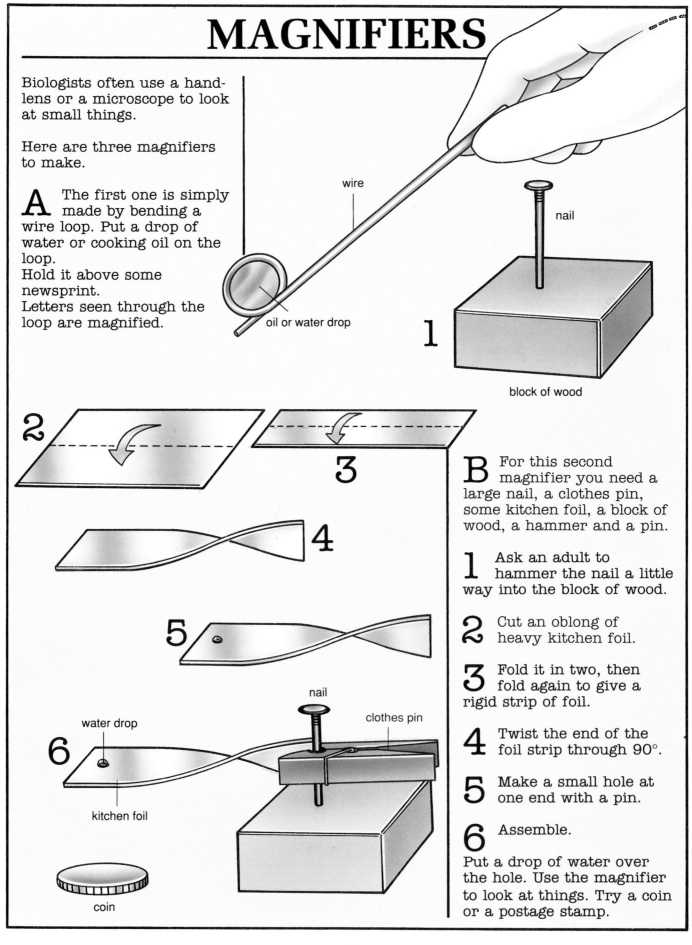

B For this second magnifier you need a large nail, a clothes pin, some kitchen foil, a block of wood, a hammer and a pin.

1 Ask an adult to hammer the nail a little way into the block of wood.

2 Cut an oblong of heavy kitchen foil.

3 Fold it in two, then fold again to give a rigid strip of foil.

4 Twist the end of the foil strip through 90°.

5 Make a small hole at one end with a pin.

6 Assemble.
Put a drop of water over the hole. Use the magnifier to look at things. Try a coin or a postage stamp.

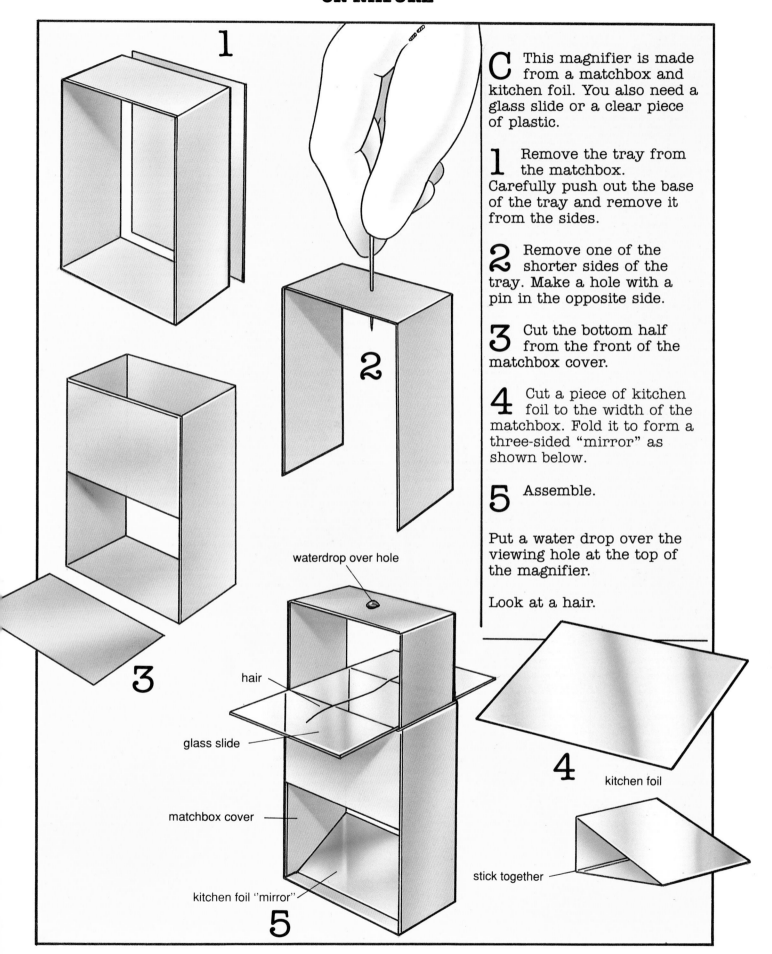

C This magnifier is made from a matchbox and kitchen foil. You also need a glass slide or a clear piece of plastic.

1 Remove the tray from the matchbox. Carefully push out the base of the tray and remove it from the sides.

2 Remove one of the shorter sides of the tray. Make a hole with a pin in the opposite side.

3 Cut the bottom half from the front of the matchbox cover.

4 Cut a piece of kitchen foil to the width of the matchbox. Fold it to form a three-sided "mirror" as shown below.

5 Assemble.

Put a water drop over the viewing hole at the top of the magnifier.

Look at a hair.

waterdrop over hole

hair

glass slide

matchbox cover

kitchen foil "mirror"

kitchen foil

stick together

NOTES FOR PARENTS AND TEACHERS

Much of the activity normally associated with nature study is of an observational kind that follows the seasons of the year. Birdwatching, country rambles and nature trails, collecting shells and so on being favorite pursuits. The favorite activities listed in this book are ones that, generally speaking, can be carried out indoors with the materials usually available in any home or school

Page 39 This draws attention to the variety of shell-life, especially if you use both univalve (single shell) and bivalve (hinged double-shell) shells to decorate the boxes.

Page 40–41 These all demonstrate a food chain. They show how animals, even carnivores, ultimately depend on plants and that a chain of organisms leads to the final consumer. In such a chain the number of organisms at each stage decreases. For example, if we take a chain that leads from plants to a hawk we would find lots of aphids feeding on the plants. A much smaller number of ladybug larvae would feed on the aphids. An even smaller number of titmice would feed on the larvae, and a few hawks feed on the titmice.

Pages 42–43 This is a simple and effective way of demonstrating four stages in the life-cycle of a butterfly. It gives an opportunity to demonstrate the meaning of items such as egg, caterpillar, chrysalis, and imago as well as metamorphosis.

Page 44 Portrait silhouettes are good for developing the concept of variation. They show how we are all alike and yet different from one another.

Page 45 Touch is a very important sense, yet looking and listening tend to dominate our lives. This game helps draw attention to its importance.

Pages 46-47 The detective work set out here brings us back to body variations, both in foot shape and size and in fingerprints. The gradual development of the concept of variation by children is important. It leads, for example, towards an eventual understanding of Darwin's theory of evolution by natural selection where nature selects from the immense variety around it.

Pages 48–49 Again the concept of variation appears for there are an infinite variety of

seeds. The idea of each seed containing an embryo plant and its food store begins to be established as children grow their seeds.

Pages 50–51 There is some emphasis on design and technology in making this miniature garden. It will be excellent if children come up with their own variations on the design!

Page 52 This helps establish that some plants grow in water.

Page 53 This activity draws attention to the fact that some plants reproduce by spores.

Pages 54–55 This establishes the hexagonal pattern of snowflakes, a theme that is taken up in "On Pattern."

Pages 56–58 These pages, while giving interesting effective records, also focus attention on form, develop appreciation of plant structure, develop manipulative skills and encourage careful observation.

Pages 59–61 These are all concerned with toys that move.
The Jumping Frog relies on energy stored in the rubber band to make it move.
The Pecking Birds work on the principle of the lever. That is to say a bar moving about a point. A *push* on the bar causes the birds to peck.
The Drumming Woodpecker has energy by virtue of its position at the top of the rod. It is said to have gravitational potential energy. When released it starts to fall. As it falls it gains the energy of movement, this is called kinetic energy.

Pages 62–63 Both these activities help develop manipulative skills as well as producing useful artefacts.

Pages 64–65 The inside of their bodies tend to be something of a mystery to children. Both models introduce children to their internal organs while at the same time being fun to make.

Pages 66–67 A considerable amount of the work of biologists deals with the microscopic world and it is probably true to say that the microscope constitutes the biologist's main tool. These pages introduce children to the idea that magnification aids are needed. Scientifically they show the magnifying property of a curved surface like that of a water drop.

3

on
Pattern

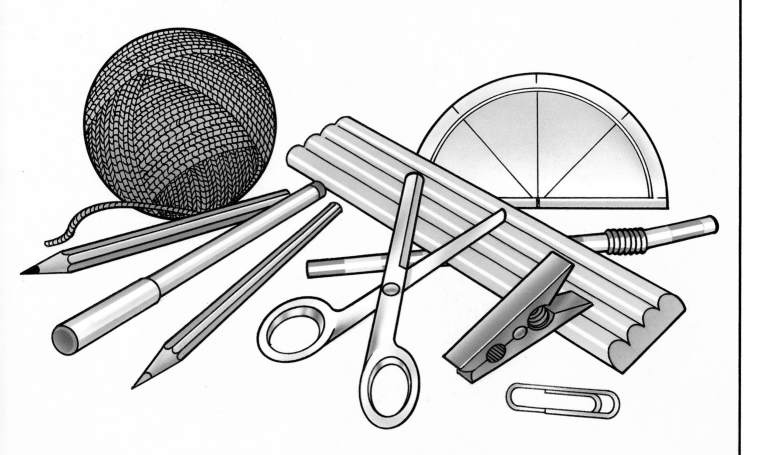

INTRODUCTION

Pattern is one of the most intriguing things in the world. Once you start to look, you will find patterns everywhere – the pattern of bricks in a wall, or tiles on a floor, or the way a mirror seems to switch the right side of your face to the left. Did you know you could make invisible patterns? You can use mirror patterns to make place mats or prints, or to make "talking" greetings cards. People adore them! You can get wonderful patterns from a swinging salt pendulum, shoot a stripy patterned aircraft through the sky, mold a patterned clay pot or make a patterned pom-pom. With a pair of compasses, a ruler and some felt-tipped pens you can create the most stunning colored designs.

Everything you need to create this world of pattern is on page 101.

Acknowledgement
I am grateful to Dr David Clinton for the design of the aircraft on Page 84

HORSES AND RIDERS

Trace the three oblongs below on to a piece of thin card. Trace the horses and jockeys too. Cut out the three oblongs. Can you find a way to place the two jockeys so that they ride the horses at a gallop? No folding allowed!

TWO INVISIBLE PATTERNS

INVISIBLE WRITING

1 Squeeze a lemon to extract the juice.

2 Take a fine paint brush and dip it in the lemon juice.

3 Paint a secret message on a piece of white paper.

4 To make the message show up you must hold it over a heat source.

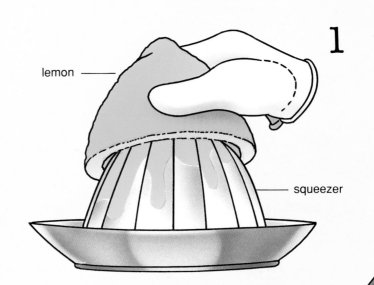

INVISIBLE PICTURES

1 Draw a picture on a white sheet of paper using the end of a candle.

2 To make the picture visible, paint all over the sheet with a weak paint mixture. The paint will not stick to the candle wax, and your drawing will show against a colored background.

ANOTHER INVISIBLE PATTERN

1 Soak a piece of writing paper in water. Use the best quality paper you can find. Leave it for a while until it feels soggy.

2 Lay the wet paper on a perfectly flat surface. A mirror is ideal. Write your initials on the soggy paper using the pointed end of an empty ballpoint pen casing.
You can draw a small design too, if you want.

3 Leave the paper to dry on the surface of the mirror so that it dries flat.
When dry it will appear blank.
If you hold it up to the light you will see that you have made your own mark in the paper. This is called a watermark.

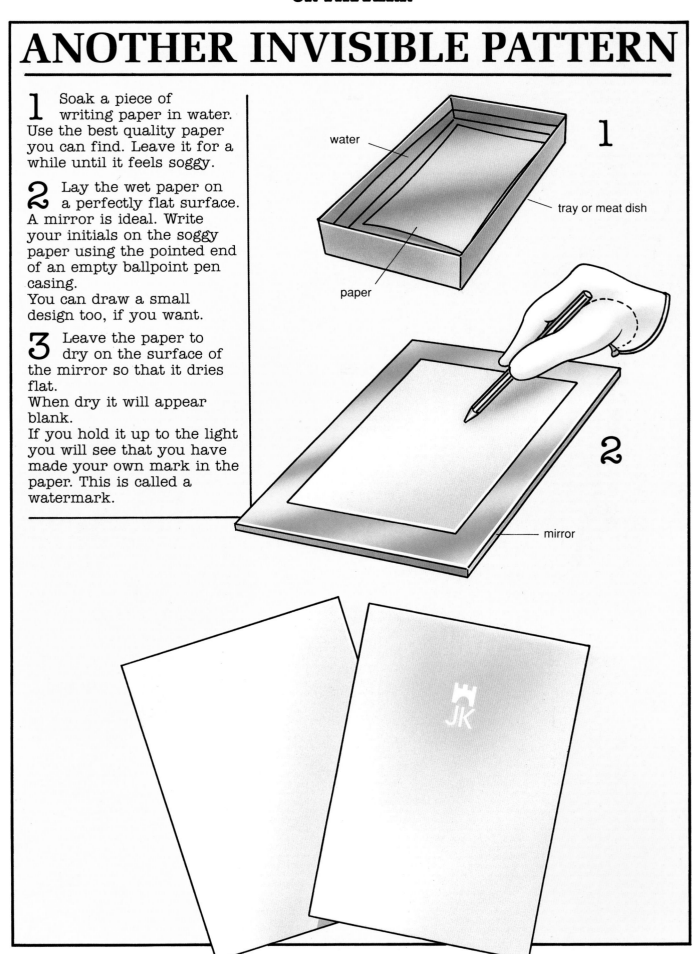

water

tray or meat dish

paper

1

mirror

2

PATTERNS FROM LETTERS

Print a letter A. Repeat it.
Repeat it again.
It makes an interesting
pattern.

Here is another pattern
using A.

Here are some more.

Try making patterns with other letters.

Here are some examples for X, Y and Z.

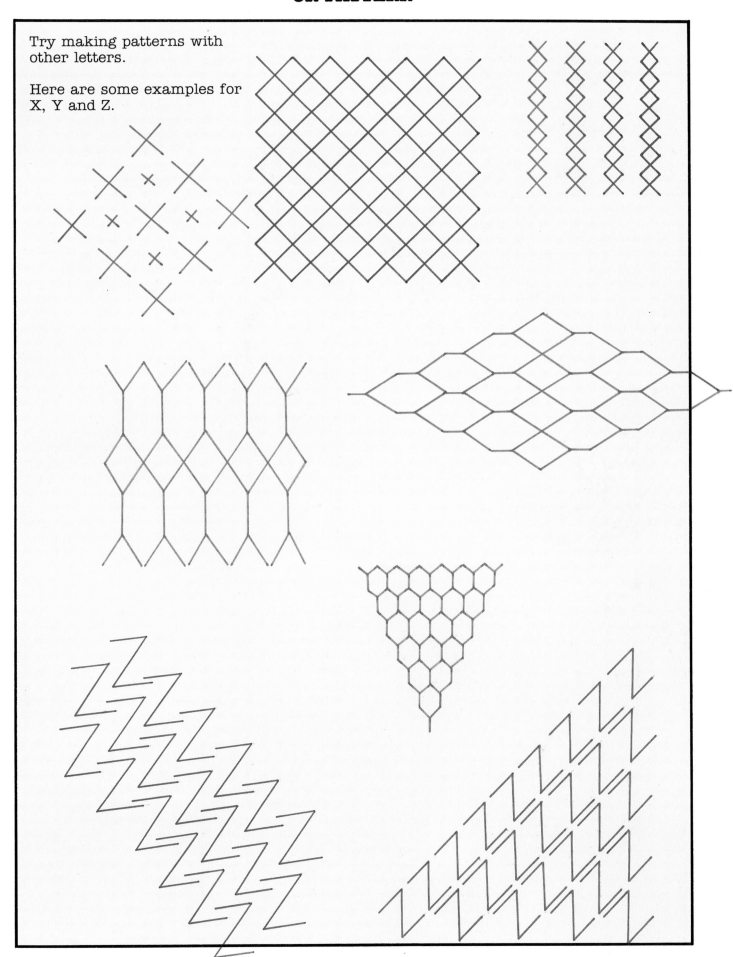

GET THE MESSAGE

Here is a card that is fun to make and to give. Follow these instructions to make a card with secret hinges. It is called a "tri-tetra flexagon" by mathematicians.

1 Trace the template on the right on to heavy paper and cut it out. Each square is $1\frac{1}{2}$ in by $1\frac{1}{2}$ in.

2 Crease the card along line B E and fold backward so that CD lies behind AB as shown.

3 Carefully bring CD forward so that it is in front of AB.

4 Fold A on top of CD. Glue it down.

5 Write a message over all of the card.

Draw a picture on the back with colored crayons.

6 Turn the card back again and fold the left side backward to lie behind the right side.

7 Separate G from H and swing G right round to finish on the right of HI.

The card now has a completely *new* back. Draw a picture on this with colored crayons. Send the card in this final jumbled form to a friend. If your friend works out the secret hingeing the pattern will become clear and the message will be revealed.

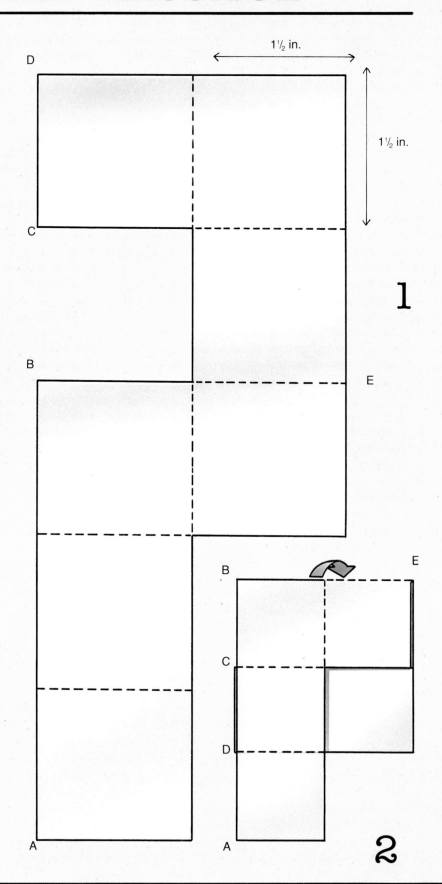

MIRROR PATTERN MATS

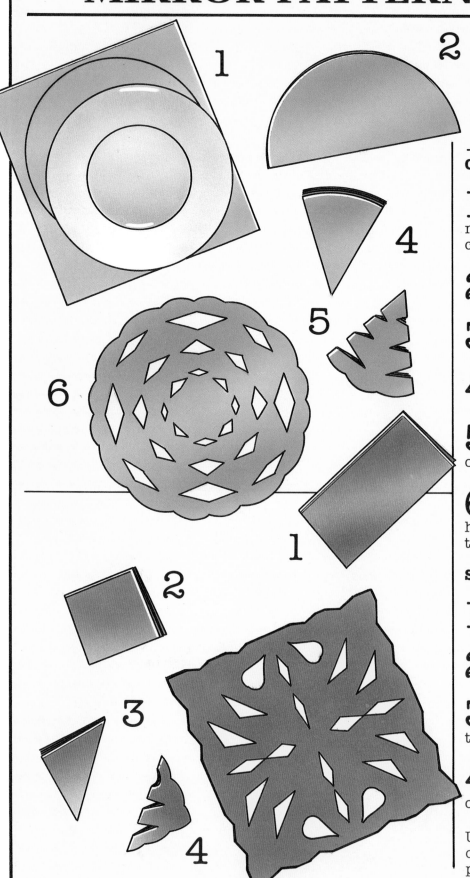

CIRCULAR MAT

1 Place a plate on a piece of paper, draw round it and cut out the circle.

2 Fold the circle in half.

3 In half again.

4 In half again.

5 Cut the edge of the circle, and snip pieces out of the folds.

6 Open the paper out to show the pattern. One half is a mirror pattern of the other half.

SQUARE MAT

1 Fold a square of paper in half.

2 Fold it in half again.

3 Fold it from corner to corner to make a triangle.

4 Make cuts along the edge and snip pieces out of the sides.

Use colored paper for colored mirror pattern place mats.

MIRROR PATTERN PRINTS

Use several newspapers piled on top of one another to make a pad.

BLOB PATTERN

1 Put a large sheet of paper on top of the pad. Draw a line down the middle. Put thick blobs of poster paint on the right hand half of the page. Keep the blobs near to the line.

2 Fold carefully in half, crease down the middle, and press evenly.

3 Open out to see the pattern.

4 Turn your pattern into a mirror pattern monster.

PAINTED PATTERN

1 Take a fresh sheet of paper. Crease it down the middle. Open it out and put it on your newspaper pad.

2 Paint half an object in one half. Paint it *thickly*.

3 Press to make the other half.

STRING PATTERN

1 Take a fresh sheet of paper and make a pattern with string dipped in paint.

2 Repeat with different colored strings.

MIRROR PATTERN BEASTS

This is a card to make for a friend who is unwell or for a birthday. It usually makes people laugh.

1 Take two sheets of paper.

Fold each piece of paper in half, short ends together, and put one piece aside. (You will need to use it later on.)

2 Mark the first piece of paper at the center of the fold.

3 Draw a line extending 2 in. from the fold.

4 Cut along this line.

5 Fold the flaps on each side of the cut back to make a triangle.

6 Fold the flaps back again and open up the page.

7 Fold the paper *up* to make a tent. Push the top triangle *down*.

Pinch the two edges of the triangle together. This will pull it through to the other side of the paper.

Do the same thing to the other triangle.
When you open and close the card it will look like a mouth talking.

8 Stick the other folded sheet to the first sheet as a backing sheet to complete the card.
Don't stick the triangles down!

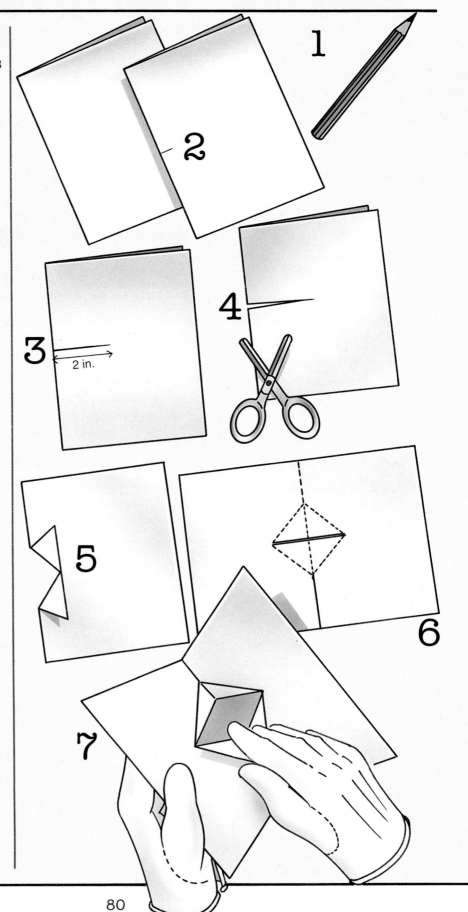

2 in.

insert and glue

Make designs around the mouth. Add your own greetings. Here are some ideas.

8

Happy Easter!

Here's Hopping

You Have A Happy Birthday!

Kiss Me!

And I Will Be Your Prince!

A POM-POM

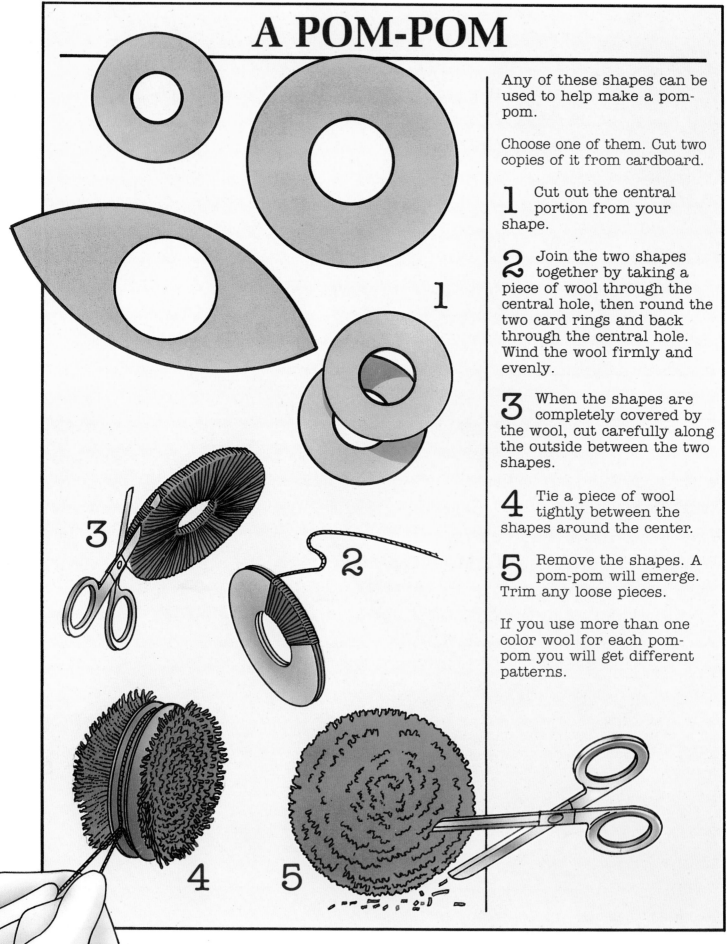

Any of these shapes can be used to help make a pom-pom.

Choose one of them. Cut two copies of it from cardboard.

1 Cut out the central portion from your shape.

2 Join the two shapes together by taking a piece of wool through the central hole, then round the two card rings and back through the central hole. Wind the wool firmly and evenly.

3 When the shapes are completely covered by the wool, cut carefully along the outside between the two shapes.

4 Tie a piece of wool tightly between the shapes around the center.

5 Remove the shapes. A pom-pom will emerge. Trim any loose pieces.

If you use more than one color wool for each pom-pom you will get different patterns.

A SALT PENDULUM

THINGS TO COLLECT

string

funnel

dispenser bottle

salt   screw eyes   nail

1 Remove the nozzle from the bottle. Keep it to use later. Cut a 1 ft. length of string.

2 Ask an adult to hammer a nail into the bottom of the bottle. Tie the string to the nail and push it right through so the bottle hangs from the string.

3 Fasten two screw eyes to the underside of an old table. (Check first with an adult before doing this!) Tie a length of string to each screw eye. Knot the other ends to the string coming from the bottle. The bottle should hang just above the floor.

4 Use a funnel to fill the bottle half full of salt. Put the nozzle back on.

5 Set the bottle swinging above a large sheet of paper.

You will get beautiful patterns.

6 Move the screw eyes to change the distance between the strings. What happens to the patterns when you do this?

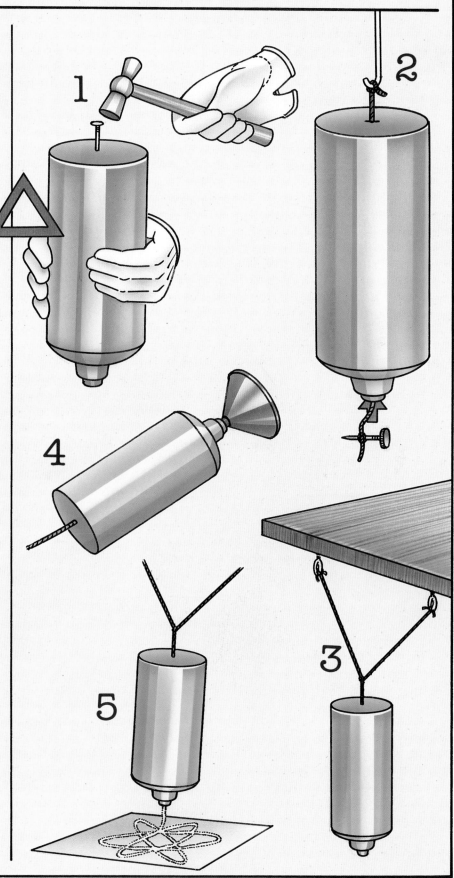

A PATTERNED PLANE

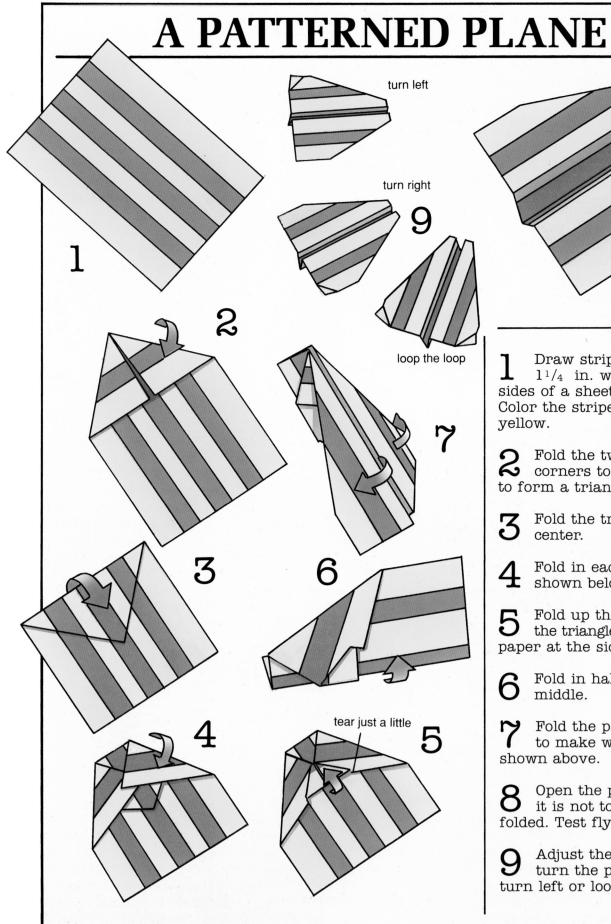

turn left

turn right

9

8

loop the loop

1

2

7

3

6

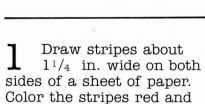

4

tear just a little

5

1 Draw stripes about 1 1/4 in. wide on both sides of a sheet of paper. Color the stripes red and yellow.

2 Fold the two top corners to the center to form a triangle.

3 Fold the triangle to the center.

4 Fold in each corner as shown below.

5 Fold up the point of the triangle, tearing the paper at the sides a little.

6 Fold in half down the middle.

7 Fold the paper down to make wings, as shown above.

8 Open the plane out so it is not too tightly folded. Test fly.

9 Adjust the wing tips to turn the plane right, turn left or loop the loop.

PATTERNED PINCH POTS

1 Roll a piece of clay about the size of a tennis ball so that it is round and smooth.

2 Push your thumb into the middle.

3 Pinch the clay between your thumb and your finger. Work around it until you have made a small bowl.

4 Pinch away to make the bowl bigger.

5 Design some patterns to go inside your bowl. Make the designs with cut pieces of clay.

6 Leave the pots to dry on a sunny windowsill.

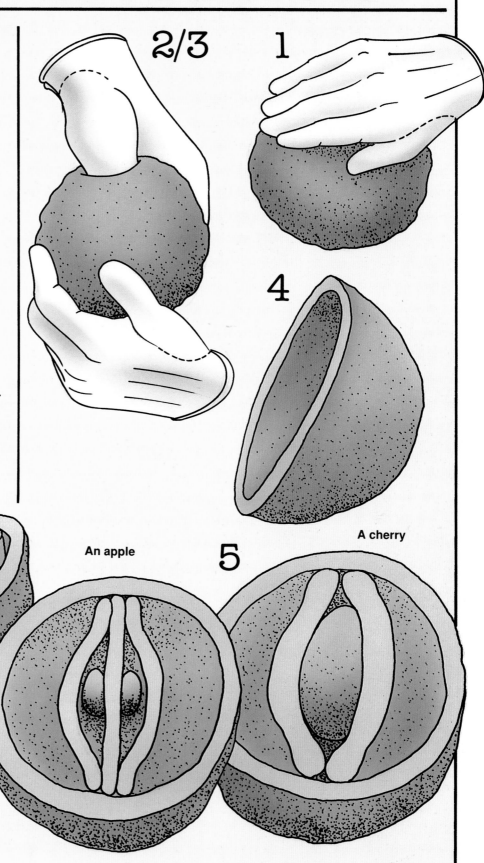

2/3

1

4

5

An orange

An apple

A cherry

LIQUID PATTERNS

Red cabbage gives a strong deep blue colored juice.

1 To get it you need to chop some red cabbage. Put it in an old saucepan with a little water.

2 Ask an adult to simmer it on the stove for five minutes.

3 Strain the liquid.

Keep this blue juice on one side.

Next collect a number of things to mix with the cabbage water.

4 Put a little of each of your ingredients into an old glass tumbler. Mix the cabbage water with each, a little at a time.

What colors do you get with different ingredients?

white wine vinegar

bicarbonate of soda

antacid drops

bath salts

lemonade

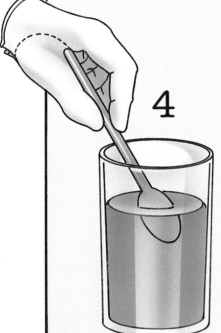

4

gives a red color when added to white wine vinegar

gives a green color when added to bicarbonate of soda

MORE LIQUID PATTERNS

You will need to be very dextrous to make this pattern!

1 Mix different amounts of salt with water as shown. Keep the same amount of water each time. Use empty plastic bottles to do the mixing in.

2 Shake the bottles every now and then until the salt dissolves. It may take two days for the largest amount of salt to dissolve!
Add a different colored food dye to three solutions.

3 Put some of the strongest solution in the bottom of an old tumbler. Carefully add some of the next strongest solution with a medicine dropper, little by little. Then the next solution and so on. If you are very, very careful you will have layers of each solution one above another.

10 tablespoons salt

2½ tablespoons salt

1

plain water

5 tablespoons salt

2

green

red

leave clear

blue

3

blue

clear

red

green

PLAYING WITH PATTERN

Each of these designs makes a pleasing pattern if it is put on to 1/2 in. grid paper.

Try it for yourself.

Here are some more regular designs for you to try. Color in your designs.

Designs that aren't regular (asymmetric designs) give even more intriguing patterns.

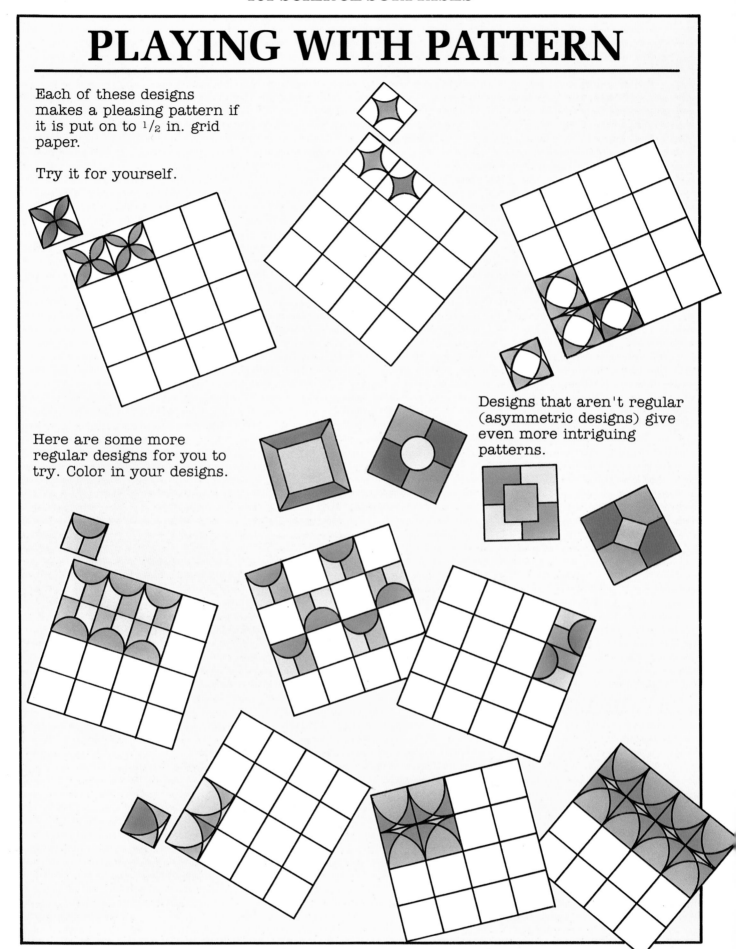

Here are more asymmetric designs to try.

With careful coloring you can make some interesting patterns. Here are some examples using this shape.

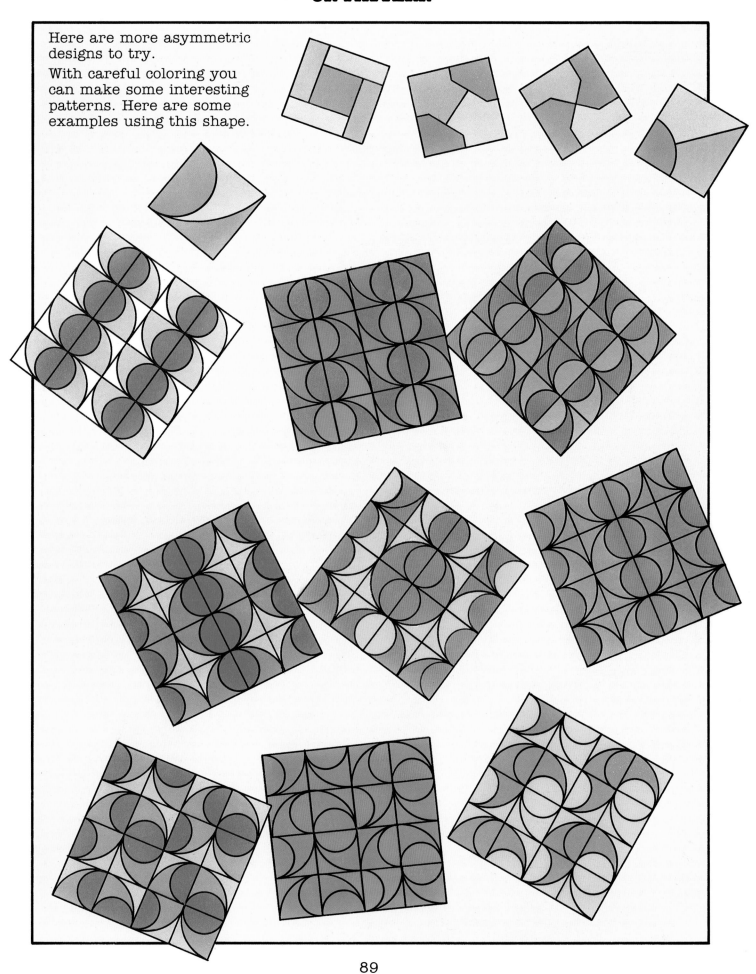

PATTERNS FROM SIX POINTS

1 Draw a circle of any radius. Mark point A on the circle.

2 Place the point of your compass on A and mark point B. Make sure you keep the distance between A and B the same as the radius of your circle.

3 Place the point of your compass on point B and mark point C. Again make sure you keep the distance between B and C the same as the radius of your circle.

4 Continue in this way round the circle. You will end up with six points each an equal distance apart.

5 With the radius of your compass kept the same draw six new circles using the six points as the centers.
Color in the pattern.

6 As you draw new circles you will find more and more points where circles cross. Drawing circles from these points gives you new patterns.

You can go on and on increasing the pattern.

PATTERNS FROM SIX POINTS

1 Draw a circle. Mark six equally spaced points as on the previous page.

2 Join the points with straight lines to make a hexagon.

3 Join alternate points to form two overlapping equal sided (equilateral) triangles.

4 Draw a smaller circle inside the new hexagon that has been formed.

5 Repeat the procedure to make reduced designs inside the smaller circle. Color your pattern.

Can you discover how to make the pattern shown above?

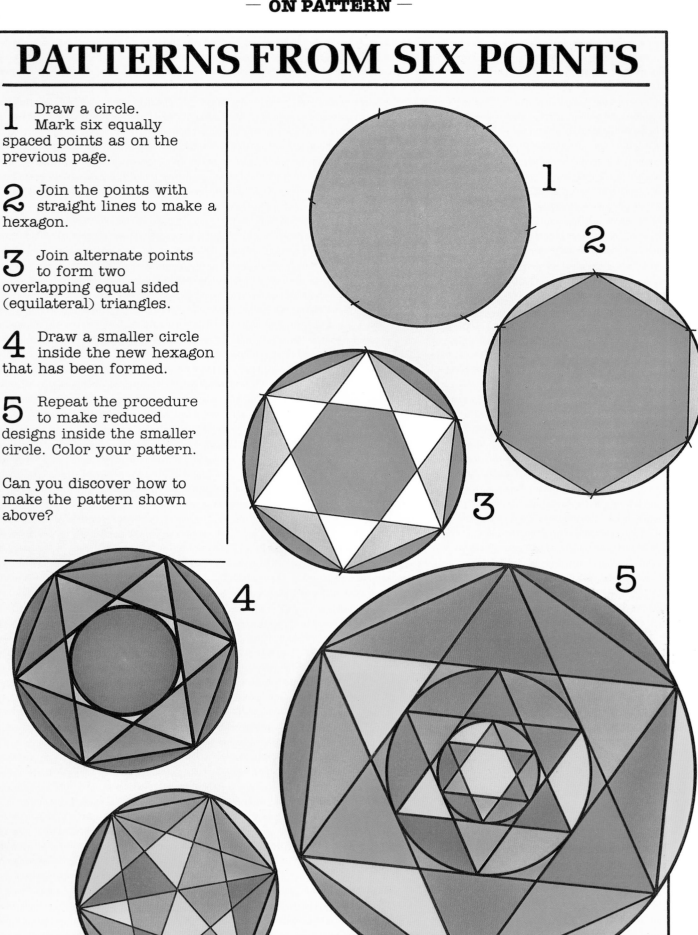

PATTERNS FROM EIGHT POINTS

1 Draw a circle. Mark off angles of 45 degrees using a protractor and a soft pencil.

2 Erase the guide lines but keep the points where the lines cross the circle. You should have 8 points.

3 Reduce the radius at which your compass is set and draw 8 arcs, one from each of the 8 points.

4 Draw a smaller circle at the center.

5 Erase any remaining guide lines and color in the pattern.

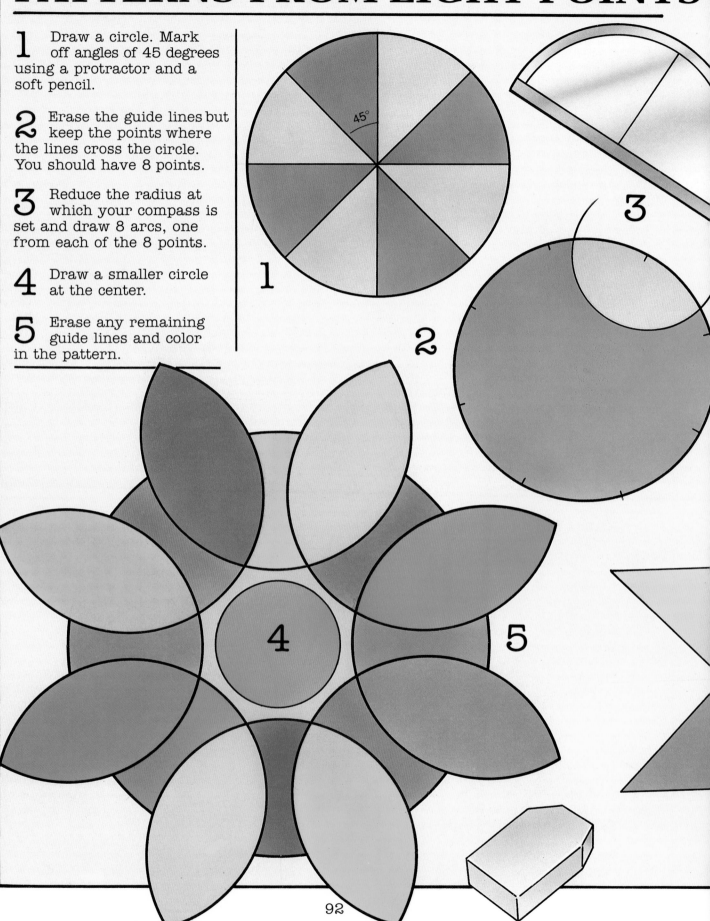

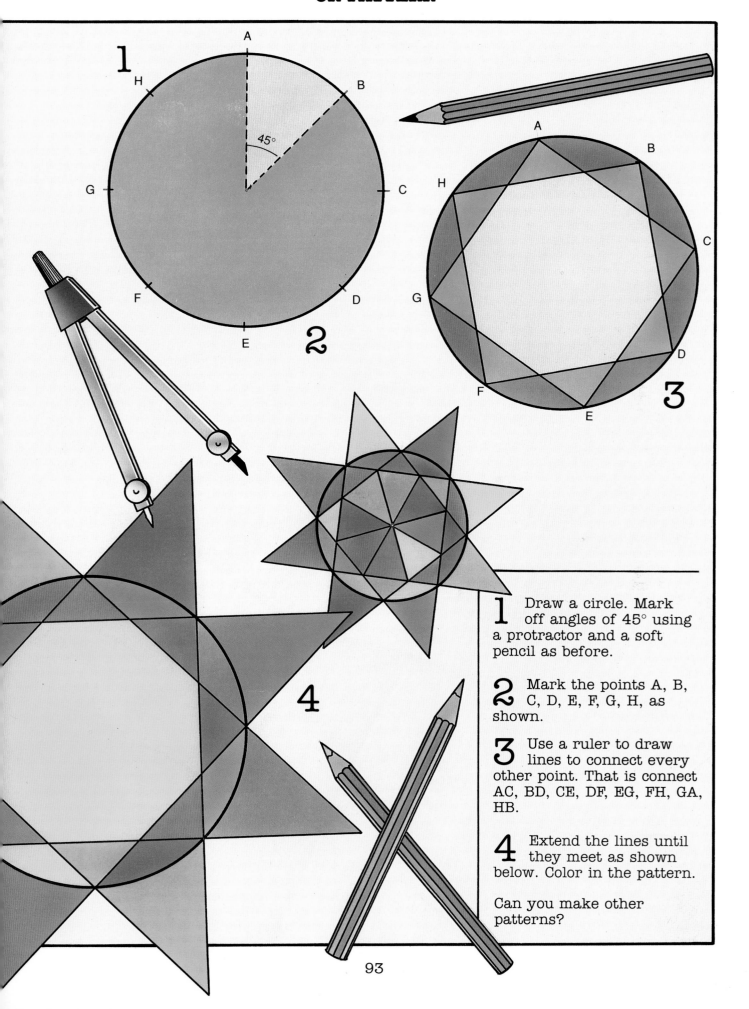

1 Draw a circle. Mark off angles of 45° using a protractor and a soft pencil as before.

2 Mark the points A, B, C, D, E, F, G, H, as shown.

3 Use a ruler to draw lines to connect every other point. That is connect AC, BD, CE, DF, EG, FH, GA, HB.

4 Extend the lines until they meet as shown below. Color in the pattern.

Can you make other patterns?

PATTERNS FROM EIGHT POINTS

Here is another intriguing eight-point pattern.

1 Draw a circle with radius ³⁄₄ in. Mark off angles of 45° using a protractor. Rule between the points made on the circle.

2 Extend each line by ³⁄₄ in.

3 Use the end point of each line as a new center to draw a ³⁄₄ in. circle.

4 Where the 8 new circles meet the original circle they cross the straight lines. Use these eight points as new centers to draw eight new circles each with radius ³⁄₄ in.

5 Color in the pattern.

Erasing some of the design and then coloring in the pattern gives new designs.

94

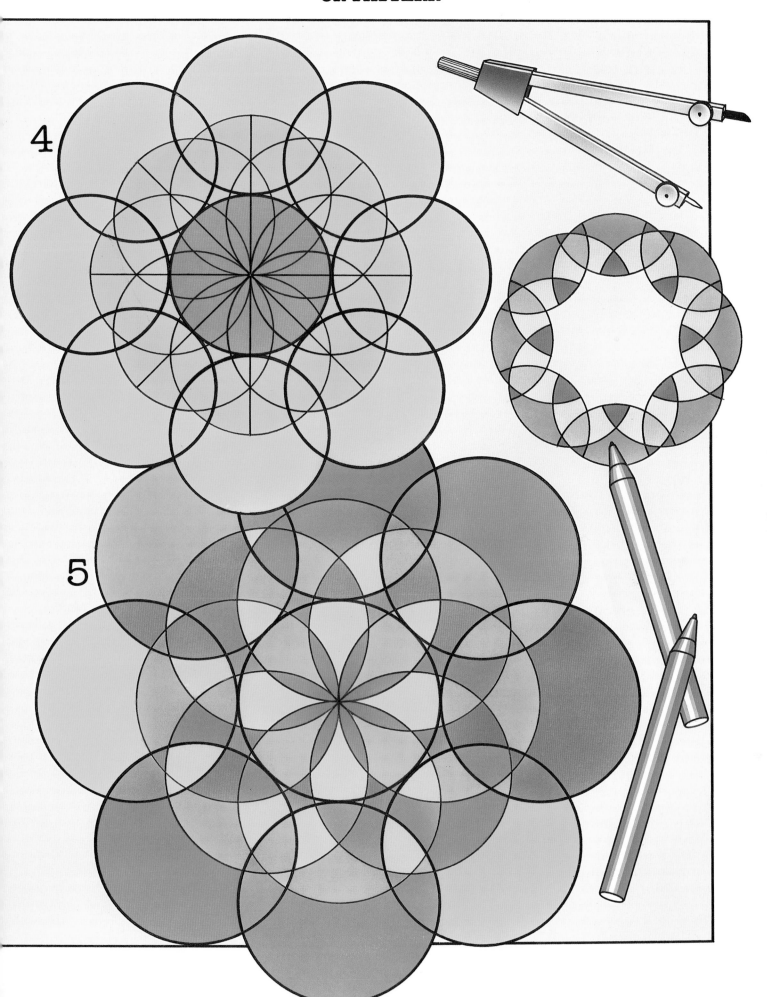

4

5

PATTERNS FROM TWELVE POINT

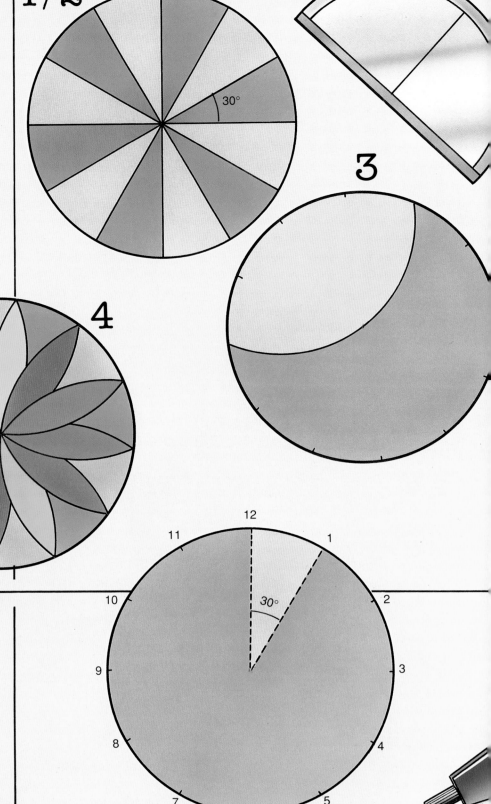

1 Draw a circle and mark off angles of 30° using a protractor and a soft pencil.

2 Erase the guide lines but keep the points where the lines cross the circle. You should have 12 points.

3 Draw an arc inside the circle from each of the 12 points with radius the same as that of the original circle.

4 Color the pattern.

Try other variations.

1 Draw a circle and mark off angles of 30° using a protractor and a soft pencil as before. Mark the points 1 to 12 as shown.

2 Keep the same radius and draw arcs inside the original circle from points 1, 2, 4, 5, 7, 8, 10, 11.

Color in the pattern.

SPIRALS

1 Draw a circle.

2 Mark off 30° angles using a protractor and a soft pencil.

3 Make a mark $1/8$ in. in on the top vertical line. Make a mark $1/4$ in. in on the next line. Mark the next line $3/8$ in. in. Continue in this way around the circle. Join the marks. You will make a spiral.

Try going in the other direction.

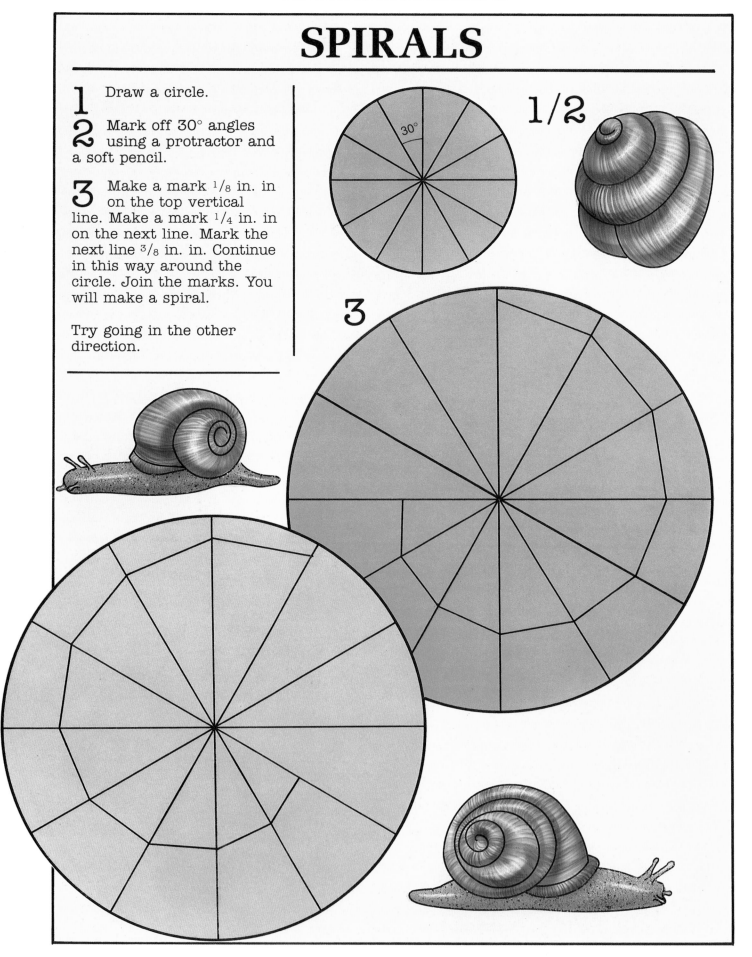

ELLIPSES

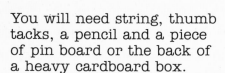

You will need string, thumb tacks, a pencil and a piece of pin board or the back of a heavy cardboard box.

1 Place a sheet of paper on your board and pin it firmly at each corner.

2 Put two thumb tacks in the board 4 in. apart as shown. Don't push them all the way in.

3 Tie a loop of fine string, longer than needed to fit around the two thumb tacks.

4 Put the loop of string around the two thumb tacks. Pull it tight with a pencil. Use the string to guide the pencil round. Make sure you keep the string taut.
You will draw an ellipse.

5 Try moving the thumb tacks close together, but keep the loop the same size.
The ellipses in the drawing shown were made by having the thumb tacks at AB, AC, AD, AE, AF, AG, AH and AI respectively.

Color in the pattern formed.

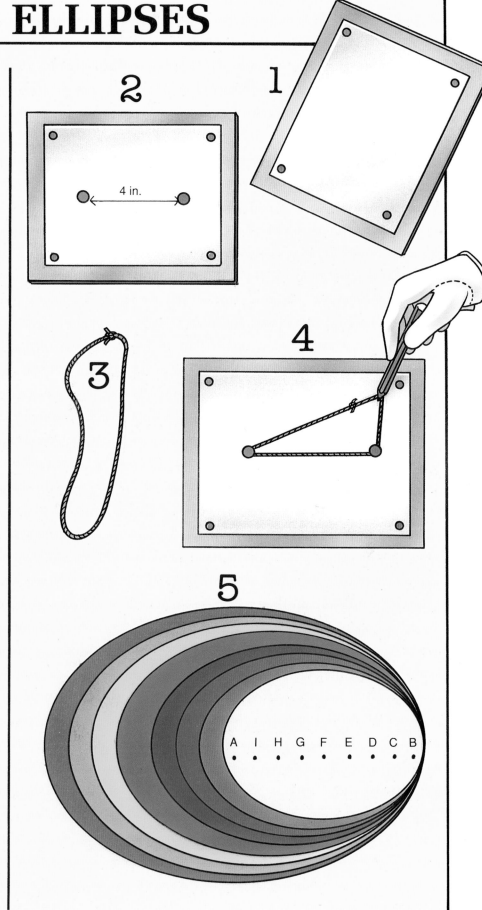

2

4 in.

3

4

5

A I H G F E D C B

NOTES FOR PARENTS AND TEACHERS

Pattern intrigues. It has infinite possibilities. This book outlines some of them, and should encourage readers to find some more.

Page 71 This puzzle was invented by the American Sam Lloyd in the last century. The circus proprietor P.T. Barnum bought thousands of these puzzles and distributed them under the title "P.T. Barnum's Trick Donkeys."

Solution

Page 72 Lemon juice breaks down on heating into a brown chemical that appears on the paper.

Page 73 Paper is made of fibers. In a cheap paper, such as newspaper, the fibers are arranged in a random pattern. In a finer quality paper the fibers tend to be much more aligned in one direction. The pressure of the point disturbs the neat arrangement of the fibers in the wet paper. When the paper dries the new order of the fibers remains. This makes the watermark.

Pages 74–75 Virtually everything can be made to show a pattern and a large part of the work of mathematicians is based on looking for pattern. On these pages it is shown how even the letters of the alphabet can be arranged to create pattern.

Pages 76–77 This card makes use of an intriguing hinging system called a tri-tetra flexagon by mathematicians.

Pages 78–79 All the activities on these pages make use of what mathematicians call reflective symmetry. That is to say, one half of a pattern reflects the other half. It is often commonly called mirror symmetry.

Pages 80–81 The cards on these pages make use of mirror symmetry. One half of each card reflects the other half. Yet another term used for this is bilateral symmetry. Biologists use this term about an animal where one half of the animal is a mirror image of the other. Both these cards take children into paper engineering and are a useful introduction to simple technology.

Page 82 Takes children into making a patterned 3D structure.

Page 83 The swinging pendulum is free to move in two planes and the salt it deposits forms the

most intriguing patterns, often similar to those obtained by drawing with a Spirograph. If the salt pattern is lightly made it can be secured to the paper by spraying with clear lacquer, although this is a task for adults.

Page 84 The patterned plane can be used to illustrate the forces of thrust, drag, lift and gravity on a flying object. This particular model moves gracefully in the air and flies long distances.

Page 85 Another 3D object to make.

Page 86 Cabbage water acts as an indicator. It will give reds, purples, blues and greens, depending on which liquid it is added to. Reds and purples indicate an acid (sour) solution while greens and blues indicate an alkaline (sweet) solution. Chemists frequently use indicators – the most well known of which must be litmus.

Page 87 The densities of liquids vary. Sea water contains a lot of salt and swimmers know well that it gives more buoyancy than fresh water. The formation of the pretty patterns of floating liquids in the activity on this page depends on some liquids being denser than others. The lighter ones float on the denser ones.

Pages 88–89 There are a number of mathematical ideas in these two pages. The initial designs at the top of Page 88 depend on "translation," that is to say the "sliding along" of a shape in one direction. The asymmetric patterns could involve this idea. They also involve such ideas as "rotation," that is the movement of a design around a fixed point rather like a clock hand rotating; or they could involve "reflection" where one design mirrors its neighbor. Using all these ideas creates a variety of interesting and dramatic patterns.

Pages 90–97 These pages are all concerned with constructing designs using a protractor and a compass.
Mathematically they bring about an understanding of angle as amount of turn, show the construction of various plane figures such as the hexagon and the octagon, as well as giving rise to a variety of patterns. There is a logic and an order to the pattern making.

Page 98 This is a useful introduction to creating spirals.

Page 99 This is an effective way of creating the curved shape which we call an ellipse. It is a useful concept to develop. Planets and satellites all move in an ellipse.

THINGS NEEDED TO MAKE THE MODELS

card
cardboard
glue
paper
pair of compasses
pens, felt tips
pencils
protractor
craft knife, pocket knife
ruler
scissors
cellophane tape
string, thin, strong

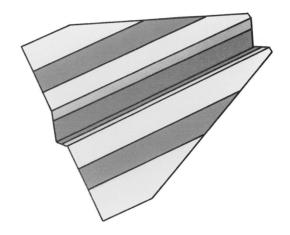

THINGS NEEDED FOR INDIVIDUAL ACTIVITIES

acid drops
balsa wood
balsa wood glue
bath salts
ballpoint, used
bicarbonate of soda
bottles, glass
bottles, plastic
bowl, glass
brush, paint
cabbage, red
candle
clay
clothespin, clip type
drinking cups, plastic
drinking straws
feathers
fishing line, nylon
hammer
jam-jar
lemon
lemon squeezer
matchsticks
medicine dropper
nails

needle, darning
newspaper
paints, poster
paper, blotting
paper clips
paper fasteners
paper-gummed
pencils
plaster of Paris
plastic wrap
pins
polystyrene block
rubber bands
salt
sandpaper (different grades)
saw, small
straw (for weaving)
thumbtacks
tongue depressers
tea
vinegar
wire, thin

INDEX